Laboratory Manual

to accompany

ELECTRONIC DEVICES AND CIRCUIT THEORY

TENTH EDITION

Robert L. Boylestad
Louis Nashelsky
Franz J. Monssen

Upper Saddle River, New Jersey
Columbus, Ohio

Editor in Chief: Vernon Anthony
Acquisitions Editor: Wyatt Morris
Editorial Assistant: Chris Reed
Project Manager: Rex Davidson
Senior Operations Supervisor: Pat Tonneman
Operations Specialist: Laura Weaver
Art Director: Diane Ernsberger
Cover Designer: Diane Ernsberger
Cover Photo: Jupiter Images
Director of Marketing: David Gesell
Marketing Assistant: Les Roberts

This book was set in New Century Schoolbook by The Special Projects Group and was printed and bound by Bind Rite Graphics. The cover was printed by Coral Graphic Services.

Copyright © 2009 by Pearson Education, Inc., Upper Saddle River, New Jersey 07458. Pearson Prentice Hall. All rights reserved. Printed in the United States of America. This publication is protected by Copyright and permission should be obtained from the publisher prior to any prohibited reproduction, storage in a retrieval system, or transmission in any form or by any means, electronic, mechanical, photocopying, recording, or likewise. For information regarding permission(s), write to: Rights and Permissions Department.

Pearson Prentice Hall™ is a trademark of Pearson Education, Inc.
Pearson® is a registered trademark of Pearson plc.
Prentice Hall® is a registered trademark of Pearson Education, Inc.

Pearson Education Ltd.
Pearson Education Singapore Pte. Ltd.
Pearson Education Canada, Ltd.
Pearson Education—Japan

Pearson Education Australia Pty. Limited
Pearson Education North Asia Ltd., Hong Kong
Pearson Educación de Mexico, S.A. de C.V.
Pearson Education Malaysia Pte. Ltd.

10 9 8 7 6 5 4 3

ISBN-13: 978-0-13-504685-2
ISBN-10:   0-13-504685-8

Contents

PREFACE		v	
EQUIPMENT LIST		vii	
EXPERIMENT	1	OSCILLOSCOPE AND FUNCTION GENERATOR OPERATION	1
EXPERIMENT	2	DIODE CHARACTERISTICS	13
EXPERIMENT	3	SERIES AND PARALLEL DIODE CONFIGURATIONS	25
EXPERIMENT	4	HALF-WAVE AND FULL-WAVE RECTIFICATION	37
EXPERIMENT	5	CLIPPING CIRCUITS	53
EXPERIMENT	6	CLAMPING CIRCUITS	67
EXPERIMENT	7	LIGHT-EMITTING AND ZENER DIODES	81
EXPERIMENT	8	BIPOLAR JUNCTION TRANSISTOR (BJT) CHARACTERISTICS	95
EXPERIMENT	9	FIXED- AND VOLTAGE-DIVIDER BIAS OF BJTS	105
EXPERIMENT	10	EMITTER AND COLLECTOR FEEDBACK BIAS OF BJTS	119
EXPERIMENT	11	DESIGN OF BJT BIAS CIRCUITS	139
EXPERIMENT	12	JFET CHARACTERISTICS	159
EXPERIMENT	13	JFET BIAS CIRCUITS	173
EXPERIMENT	14	DESIGN OF JFET BIAS CIRCUITS	187
EXPERIMENT	15	COMPOUND CONFIGURATIONS	203
EXPERIMENT	16	MEASUREMENT TECHNIQUES	221
EXPERIMENT	17	COMMON-EMITTER TRANSISTOR AMPLIFIERS	239
EXPERIMENT	18	COMMON-BASE AND EMITTER-FOLLOWER (COMMON-COLLECTOR) TRANSISTOR AMPLIFIERS	249
EXPERIMENT	19	DESIGN OF COMMON-EMITTER AMPLIFIERS	269
EXPERIMENT	20	COMMON-SOURCE TRANSISTOR AMPLIFIERS	277
EXPERIMENT	21	MULTISTAGE AMPLIFIERS: RC COUPLING	287
EXPERIMENT	22	CMOS CIRCUITS	303
EXPERIMENT	23	DARLINGTON AND CASCODE AMPLIFIER CIRCUITS	309

EXPERIMENT	24	CURRENT SOURCE AND CURRENT MIRROR CIRCUITS	321
EXPERIMENT	25	FREQUENCY RESPONSE OF COMMON-EMITTER AMPLIFIERS	329
EXPERIMENT	26	CLASS-A AND CLASS-B POWER AMPLIFIERS	339
EXPERIMENT	27	DIFFERENTIAL AMPLIFIER CIRCUITS	351
EXPERIMENT	28	OP-AMP CHARACTERISTICS	369
EXPERIMENT	29	LINEAR OP-AMP CIRCUITS	377
EXPERIMENT	30	ACTIVE FILTER CIRCUITS	389
EXPERIMENT	31	COMPARATOR CIRCUITS OPERATION	401
EXPERIMENT	32	OSCILLATOR CIRCUITS 1: THE PHASE-SHIFT OSCILLATOR	411
EXPERIMENT	33	OSCILLATOR CIRCUITS 2	419
EXPERIMENT	34	VOLTAGE REGULATION—POWER SUPPLIES	429
EXPERIMENT	35	ANALYSIS OF AND, NAND, AND INVERTER LOGIC GATES	437
EXPERIMENT	36	ANALYSIS OF OR, NOR, AND XOR LOGIC GATES	449
EXPERIMENT	37	ANALYSIS OF INTEGRATED CIRCUITS	461

Preface

For this edition our effort was directed toward ensuring that the data obtained was meaningful and the instructions clear. Each experiment has now been class tested for the past three years and should be in good form. Although some experiments reflect suggestions submitted by reviewers, the heading for each remains the same.

The first half of the manual is devoted primarily to DC analysis of electronic circuits, with the second half devoted to AC operation of electronic circuits.

Three experiments for logic circuits that contain simple logic gates and integrated circuits that contain flip-flops have been added to this edition. Such devices and circuits play an increasing role in developing electronic technology and are important for engineering students and practicing professionals to understand. These experiments stress the close approximation between computer simulation and the experimental analysis and design of such circuits. Such an approximation is made possible by the ever-increasing sophistication and proliferation of computer technology.

Each of the laboratory experiments has PSpice simulations that relate directly to these laboratory experiments. It is strongly recommended that students perform the simulations before they do the actual experiment in the laboratory. The simulation data become the template against which they can compare their experimental results. Thus, they discover any discrepancies between simulation data and experimental results in a timely fashion. Needed corrections of simulation data, laboratory procedure, and data collection can be performed during the course of an experiment.

In all the experiments, graphs are provided for plotting data or recording waveforms. In addition, space is set aside for calculations and to answer questions. When the experiment is completed, the student can remove the page from the laboratory manual at the perforations and submit it for grading. There should seldom be a need for additional pages to report on the experiment.

An equipment list is provided to help the student determine the availability of equipment to run all the laboratory experiments. A sincere effort was made to limit the parts list and to keep the power levels as low as possible.

The authors wish to extend their sincerest appreciation to Professors Bill Boettcher, Jake Froese, Doug Fuller, Lee Rosenthal, and Gerald Terrebrood for their excellent reviews and very helpful suggestions.

Robert Boylestad
Louis Nashelsky
Franz Monssen

Equipment List

EXPERIMENTS 1-15

INSTRUMENTS

Oscilloscope (dual trace preferable)
Digital Multimeter (DMM)
DC Power Supply
Signal Generator
Frequency Counter

RESISTORS*

(1) 100 Ω   (1) 2 kΩ   (2) 15 kΩ
(1) 220 Ω   (1) 2.2 kΩ   (1) 33 kΩ
(1) 300 Ω   (1) 2.4 kΩ   (1) 100 kΩ
(1) 330 Ω   (1) 2.7 kΩ   (1) 330 kΩ
(1) 470 Ω   (1) 3 kΩ   (1) 390 kΩ
(1) 680 Ω   (1) 3.3 kΩ   (1) 1 MΩ
(2) 1 kΩ   (1) 3.9 kΩ   (1) 10 MΩ
(1) 1.2 kΩ   (1) 4.7 kΩ
(1) 1.5 kΩ   (1) 6.8 kΩ
(1) 1.8 kΩ   (1) 10 kΩ

(1) 1 kΩ potentiometer
(1) 5 kΩ potentiometer
(1) 1 MΩ potentiometer

CAPACITORS

(1) 0.1 µF
(1) 1 µF

DIODES

(4) Silicon
(1) Germanium
(1) LED
(1) Zener (10 V)

---

*All resistors 1/2 W, unless otherwise indicated.

TRANSISTORS

(2) BJT 2N3904 (or equivalent)
(1) BJT 2N4401 (or equivalent)
(1) JFET 2N4416 (or equivalent)
(1) BJT without terminal identification

MISCELLANEOUS

(1) 12.6 V Center-tapped transformer with fused line cord
(1) Heat gun (if available)
(1) Curve tracer (if available)

EXPERIMENTS 16-37

INSTRUMENTS

Oscilloscope (dual trace preferable)
Digital Multimeter (DMM)
DC Power Supply
Signal Generator
Frequency Counter
Cadet Logic Trainer (or equivalent)
Computer with OrCad PSpice version 9.1 or higher

RESISTORS*

(1) 20 Ω	(2) 1 kΩ	(1) 5.1 kΩ	(1) 220 kΩ
(1) 51 Ω, 1 W	(2) 1.2 kΩ	(1) 5.6 kΩ	(2) 1 MΩ
(1) 82 Ω	(1) 1.8 kΩ	(1) 6.8 kΩ	
(2) 100 Ω	(1) 2 kΩ	(1) 7.5 kΩ	
(1) 120 Ω, 0.5 W	(2) 2.2 kΩ	(5) 10 kΩ	
(1) 150 Ω	(2) 2.4 kΩ	(2) 20 kΩ	
(1) 180 Ω	(2) 3 kΩ	(1) 27 kΩ	
(2) 390 Ω, 0.5 W	(1) 3.3 kΩ	(1) 33 kΩ	
(1) 510 Ω	(1) 3.9 kΩ	(1) 39 kΩ	
(1) 1 kΩ, 0.5 W	(1) 4.3 kΩ	(1) 51 kΩ	
(1) 5 kΩ pot.	(1) 4.7 kΩ	(3) 100 kΩ	

(1) 50 kΩ potentiometer
(1) 500 kΩ potentiometer

CAPACITORS

(3) 0.001 μF
(3) 0.01 μF
(3) 0.1 μF
(1) 1 μF
(4) 10 μF
(3) 15 μF
(2) 20 μF
(2) 100 μF

---

*All resistors 1/2 W, unless otherwise indicated.

DIODES

(2) Silicon
(1) LED (20 mA)

TRANSISTORS

(3) BJT 2N3904 (or equivalent)
(3) JFET 2N3823 (or equivalent)
(1) TIP 120
(1) 2N4300 npn medium power
(1) 2N5333 pnp medium power

ICs

(1) 74HC02 or 14002 CMOS gate
(1) 74HC04 or 14004 CMOS inverter
(1) 7414 Schmitt-trigger hex inverter
(1) 301 op-amp
(1) 339 comparator
(1) 741 op-amp
(1) 7400
(1) 7402

(1) 7404
(2) 7408
(1) 7432
(1) 7474
(1) 7486
(1) 7493A
(2) 74107

ALL EXPERIMENTS

INSTRUMENTS

Oscilloscope (dual trace preferable)
Digital Multimeter (DMM)
DC Power Supply
Signal Generator
Frequency Counter

RESISTORS*

(1) 20 Ω	(2) 1 kΩ	(1) 6.8 kΩ
(1) 51 Ω, 1 W	(2) 1 kΩ, 0.5 W	(1) 7.5 kΩ
(1) 82 Ω	(2) 1.2 kΩ	(5) 10 kΩ
(2) 100 Ω	(1) 1.8 kΩ	(2) 15 kΩ
(1) 120 Ω, 0.5 W	(1) 2 kΩ	(2) 20 kΩ
(1) 150 Ω	(2) 2.2 kΩ	(1) 27 kΩ
(1) 180 Ω	(2) 2.4 kΩ	(1) 33 kΩ
(1) 220 Ω	(1) 2.7 kΩ	(1) 39 kΩ
(1) 300 Ω	(2) 3 kΩ	(1) 51 kΩ
(1) 330 kΩ	(1) 3.3 kΩ	(3) 100 kΩ
(1) 390 Ω, 0.5 W	(1) 3.9 kΩ	(1) 220 kΩ
(1) 470 Ω	(1) 4.3 kΩ	(1) 330 kΩ
(2) 510 Ω	(1) 4.7 kΩ	(1) 390 kΩ
(1) 680 Ω	(5) 5.1 kΩ	(2) 1 MΩ
	(1) 5.6 kΩ	(1) 10 MΩ

*All resistors 1/2 W, unless otherwise indicated.

(1) 1 kΩ potentiometer
(1) 5 kΩ potentiometer
(1) 50 kΩ potentiometer
(1) 500 kΩ potentiometer
(1) 1 MΩ potentiometer

CAPACITORS

(3) 0.001 μF
(3) 0.01 μF
(3) 0.1 μF
(1) 1 μF
(4) 10 μF
(3) 15 μF
(2) 20 μF
(2) 100 μF

DIODES

(4) Silicon
(1) Germanium
(1) LED (20 mA)
(1) Zener (10 V)

TRANSISTORS

(3) BJT 2N3904 (or equivalent)
(3) JFET 2N3823 (or equivalent)
(1) BJT 2N4401 (or equivalent)
(1) JFET 2N4416 (or equivalent)
(1) TIP 120
(1) 2N4300 npn medium power
(1) 2N5333 pnp medium power
(1) BJT without terminal identification

ICs

(1) 74HC02 or 14002 CMOS gate
(1) 74HC04 or 14004 CMOS inverter
(1) 7414 Schmitt-trigger hex inverter
(1) 301 op-amp
(1) 339 comparator
(1) 741 op-amp

MISCELLANEOUS

(1) 12.6 V Center-tapped transformer with fused line cord
(1) Heat gun (if available)
(1) Curve tracer (if available)

NOTICE TO READER

The publisher and the author(s) do not warrant or guarantee any of the products and/or equipment described herein nor has the publisher or the author(s) made any independent analysis in connection with any of the products, equipment, or information used herein. The reader is directed to the manufacturer for any warranty or guarantee for any claim, loss, damages, costs or expense, arising out of or incurred by the reader in connection with the use or operation of any of the products and/or equipment.

No experiment should be undertaken without proper supervision, equipment and safety precautions and all accepted safety standards and warnings. DO NOT attempt to perform these experiments relying solely on the information presented in this manual. The reader is expressly advised to adopt all safety precautions that might be indicated in the products and/or equipment by the activities and experiments described herein. The reader assumes all risks in connection with such instructions.

Name _____
Date _____
Instructor _____

EXPERIMENT 1

Oscilloscope and Function Generator Operation

OBJECTIVE

To use the oscilloscope and function generator to calculate, obtain, and measure the amplitudes and durations (periods) of various voltage signals.

EQUIPMENT REQUIRED

Instruments

Oscilloscope
Digital multimeter

Supplies

(1) 1.5-V D cell and holder
Function generator

EQUIPMENT ISSUED

Item	Laboratory serial no.
Oscilloscope	
DMM	
Function generator	

RÉSUMÉ OF THEORY

Oscilloscope

The oscilloscope is the most important instrument available to the practicing technician or engineer. It permits the visual display of a voltage signal that can reveal a range of information of the operating characteristics of a circuit or system that is not available with a standard multimeter. At first glance the instrument may appear complex and difficult to master. Be assured, however, that once the function of each section of the oscilloscope is explained and understood and the system is used throughout a set of experiments, your expertise with this important tool will develop quite rapidly.

In addition to the display of a signal, it can also be used to measure the average value, rms value, frequency, and period of a sinusoidal or nonsinusoidal signal. The screen is divided into centimeter divisions in the vertical and horizontal directions. The vertical sensitivity is provided (or set) in volts/div., while the horizontal scale is provided (or set) in t time (s/div.). If a particular signal occupies 6 vertical divisions and the vertical sensitivity is 5 mV/div., the magnitude of the signal can be determined from the following equation:

Amplitude of signal voltage = voltage sensitivity (V/div.) × deflection (div.)
$$V_s = (5 \text{ mV/div.})(6 \text{ div.}) = 30 \text{ mV} \tag{1.1}$$

If one cycle of the same signal occupies 8 divisions on the horizontal scale with a horizontal sensitivity of 5 µs/div., the period and frequency of the signal can be determined using the following equations:

Period of signal voltage = horizontal sensitivity (s/div.) × deflection (div.)
$$T = (5 \text{ µs/div.})(8 \text{ div.}) = 40 \text{ µs} \tag{1.2}$$

$$\text{and } f = \frac{1}{T} = \frac{1}{40 \text{ µs}} = 25 \text{ kHz}$$

Function Generator

The function generator is a voltage supply that typically provides a sinusoidal, square-wave, and triangular voltage waveform for a range of frequencies and amplitudes. The frequency and the amplitude of these voltage functions can be set by the proper dial positions and their associated multipliers. For more precise settings of these parameters, the oscilloscope is used.

Both the scope and the function generator are built to withstand some abuse, so do not be afraid to try various combinations of dial settings to fully develop your abilities with this laboratory experiment. If you are working in a group, every group member should be involved in the experimental work. It is important to learn how to use the laboratory equipment, such as the function generator and the oscilloscope, properly. Such acquired skills are essential for the job of the electrical engineer and technician.

Exp. 1 / Procedure

PROCEDURE

Part 1. The Oscilloscope

The instructor will provide a brief description of the various sections of the oscilloscope and function generator. In your own words, describe the function and use of each of the following controls or sections of the oscilloscope.

 a. Focus of voltage trace:

 b. Intensity of voltage trace:

 c. Vertical and horizontal position controls:

 d. Vertical sensitivity:

 e. Horizontal sensitivity:

 f. Vertical mode selection:

 g. AC-GND-DC switch:

 h. Beam finder:

 i. Calibrate switches:

 j. Internal sync:

 k. Trigger section:

 l. External trigger input:

m. Input resistance and capacitance:

n. Probe:

Part 2. The Function Generator

Setup

a. Turn on the oscilloscope and adjust the necessary controls to establish a clear, bright, horizontal line across the center of the screen. Adjust the various controls to see their effects on the display.

b. Connect the function generator to one vertical channel of the oscilloscope and set the output of the generator to a 1000 Hz sinusoidal waveform.

c. Set the vertical sensitivity of the scope to 1 V/div. and adjust the amplitude control of the function generator to establish a 4 V peak-to-peak (p-p) sinusoidal waveform on the screen.

Horizontal Sensitivity

d. Determine the period of the 1000 Hz sinusoidal waveform in milliseconds using the equation $T = 1/f$. Show all work for each part of the experiment. Be neat!

T (calculated) = _____

e. Set the horizontal sensitivity of the scope to 0.2 ms/div. Using the results of Part 2(**d**), predict and calculate the number of horizontal divisions required to properly display one full cycle of the 1000 Hz signal.

Number of divisions (calculated) = _____

Using the oscilloscope, measure the number of required divisions and insert below. How does the result compare to the calculated number of divisions?

Number of divisions (measured) = _____

f. Change the horizontal sensitivity of the oscilloscope to 0.5 ms/div. without touching any of the controls of the function generator. Using the results of Part 2(**d**), how many horizontal divisions will now be required to display one full cycle of the 1000 Hz signal?

Number of divisions (calculated) = _____

Using the oscilloscope, measure the number of required divisions and insert below. How does the result compare to the calculated number of divisions?

Number of divisions (measured) = _____

g. Change the horizontal sensitivity of the oscilloscope to 1 ms/div. without touching any of the controls of the function generator. Using the results of Part 2(**d**), how many horizontal divisions will now be required to display one full cycle of the 1000 Hz signal?

Number of divisions (calculated) = _____

Using the oscilloscope, measure the number of required divisions and insert below. How does the result compare to the calculated number of divisions?

Number of divisions (measured) = _____

h. What was the effect on the appearance of the sinusoidal waveform as the horizontal sensitivity was changed from 0.2 ms/div. to 0.5 ms/div. and finally to 1 ms/div.?

Did the frequency of the signal on the screen change with each horizontal sensitivity? What conclusion can you draw from the results regarding the effect of the chosen horizontal sensitivity on the signal output of the function generator?

i. Given a sinusoidal waveform on the screen, review the procedure to determine its frequency. Develop a sequence of steps to calculate the frequency of a sinusoidal waveform appearing on the screen of an oscilloscope.

Vertical Sensitivity

j. Do not touch the controls of the function generator but set the sensitivity of the scope to 0.2 ms/div. and set the vertical sensitivity to 2 V/div. Using this latter sensitivity, calculate the peak-to-peak value of the sinusoidal waveform on the screen by first counting the number of vertical divisions between peak values and multiplying by the vertical sensitivity.

Peak-to-peak value (calculated) = _____

k. Change the vertical sensitivity of the oscilloscope to 0.5 V/div. and repeat Part 2(j).

Peak-to-peak value (calculated) = _____

l. What was the effect on the appearance of the sinusoidal waveform as the vertical sensitivity was changed from 2 V/div. to 0.5 V/div.?

Did the peak-to-peak voltage of the sinusoidal signal change with each vertical sensitivity? What conclusion can you draw from the results regarding the effect of changing the vertical sensitivity on the output signal of the function generator?

m. Can the peak or peak-to-peak output voltage of a function generator be set without the aid of an auxiliary instrument such as an oscilloscope or DMM? Explain.

Part 3. Exercises

a. Make all the necessary adjustments to clearly display a 5000-Hz 6 V_{p-p} sinusoidal signal on the oscilloscope. Establish the zero volt line at the center of the screen.

Record the chosen sensitivities:

Vertical sensitivity = _____

Horizontal sensitivity = _____

Draw the waveform on Fig. 1.1, carefully noting the required number of horizontal and vertical divisions. Add vertical and horizontal dimensions to the waveform using the chosen sensitivities listed above.

Exp. 1 / Procedure

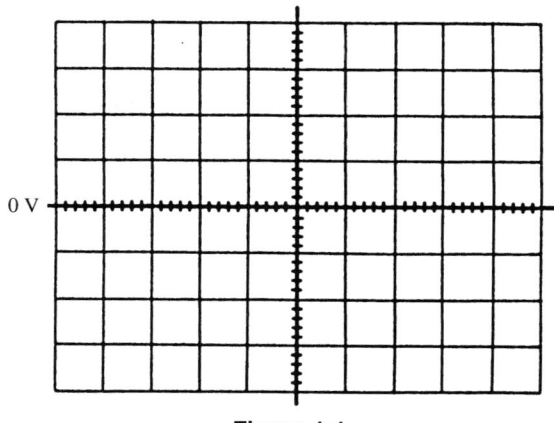

Figure 1-1

Calculate the period of the waveform on the screen using the number of horizontal divisions for a full cycle as shown.

T (calculated) = _____

b. Repeat Part 3(**a**) for a 200-Hz 0.8 V_{p-p} sinusoidal waveform on Fig. 1.2.

Vertical sensitivity = _____
Horizontal sensitivity = _____

T (calculated) = _____

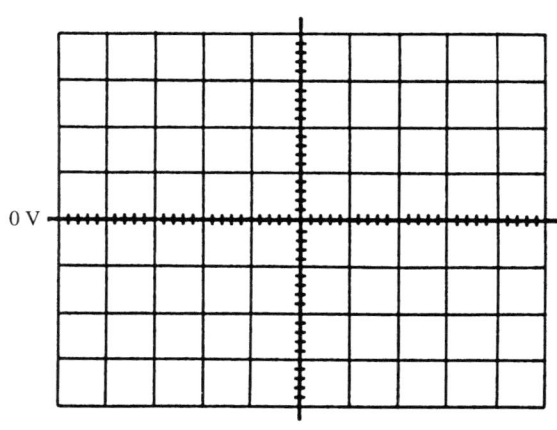

Figure 1-2

c. Repeat Part 3(**a**) for a 100-kHz 4 V_{p-p} square wave on Fig. 1.3. Note that a square wave is called for.

Vertical sensitivity = _____
Horizontal sensitivity = _____

T (calculated) = _____

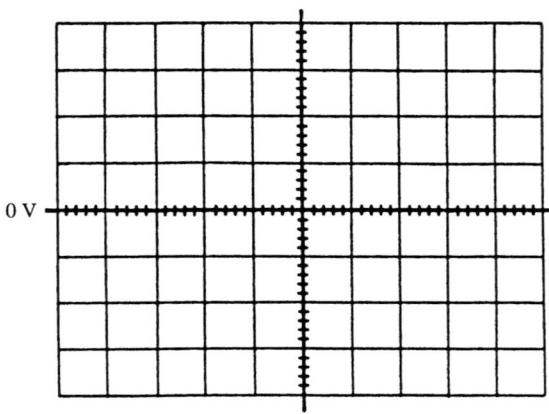

Figure 1-3

Part 4. Effect of DC Levels

a. Reestablish the 1-kHz 4 V_{p-p} sinusoidal waveform on the screen. Calculate the effective value of the sinusoidal waveform.

V_{rms} (calculated) = _____

b. Disconnect the function generator from the scope and measure the effective (rms) value of the output of the function generator using the digital meter.

V_{rms} (measured) = _____

c. Determine the magnitude of the percent difference between the calculated and measured levels using the following equation:

$$\% \text{ Difference} = \left| \frac{V_{(calc)} - V_{(meas)}}{V_{(calc)}} \right| \times 100\%$$

% Difference = _____

d. Reconnect the function generator to the scope with the 1-kHz 4 V_{p-p} signal and switch the AC-GND-DC coupling switch of the vertical channel to GND. What is the effect?

Why? How can this scope function be used?

e. Now move the AC-GND-DC coupling switch to the AC position. What is the effect on the screen display? Why?

f. Finally, move the AC-GND-DC coupling switch to the DC position. What is the effect on the screen display (if any)? Why?

g. Construct the input of v_i of Fig. 1.4 by placing a D cell in series with the output of the function generator. Be sure the ground of the oscilloscope is connected directly to the ground of the function generator. Measure and record the actual battery voltage using the DC mode of the DMM.

DC level (measured) = _____

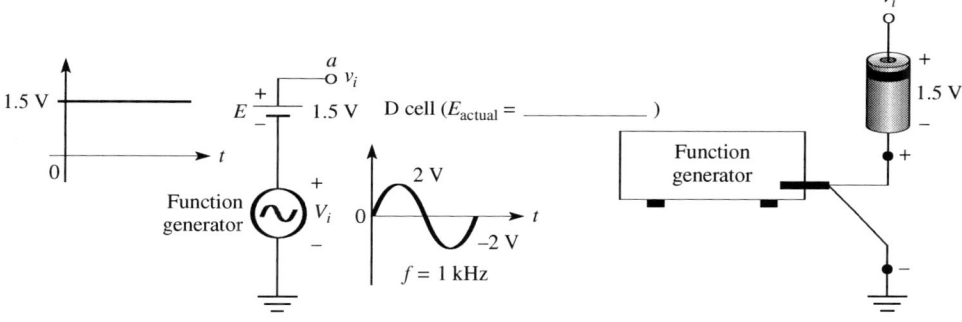

Figure 1-4

h. Apply the input voltage v_i of Fig. 1.4 to one channel of the oscilloscope with the AC-GND-DC coupling switch in the GND position and set the resulting horizontal line (zero reference level) in the middle of the screen. Then move the AC-GND-DC coupling switch to the AC position and make a rough sketch of the waveform on Fig. 1.5, clearly showing the zero reference line and the number of vertical and horizontal divisions. Using the chosen sensitivities, label the magnitudes of the various horizontal and vertical grid lines.

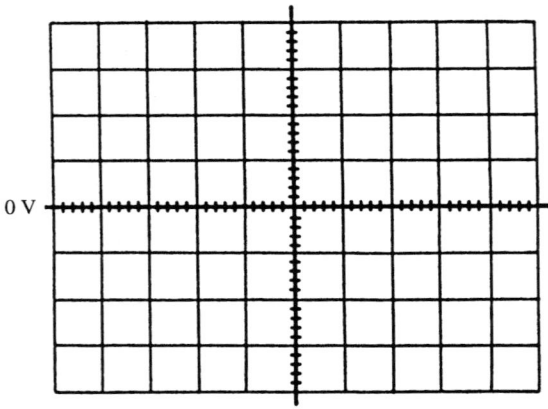

Figure 1-5

i. Switch the position of the AC-GND-DC coupling switch to the DC mode and make a rough sketch of the resulting waveform on Fig. 1.6, including the detail requested in Part 4(**h**).

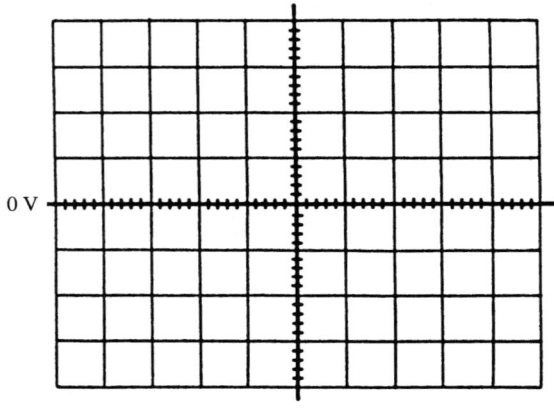

Figure 1-6

Did the vertical shift of the sinusoidal waveform equal the DC voltage of the battery?

Is the shape of the sinusoidal waveform changed by moving the AC-GND-DC coupling switch through the various positions?

j. Reverse the polarity of the battery of Fig. 1.4 and repeat Parts 4(**h**) and (**i**). Observe the effect on the waveform in the AC and DC modes and comment below.

Part 5. Problems

1. Given $v = 20 \sin 2000t$, determine
   a. ω

   b. f

   c. T

   d. Peak value

   e. Peak-to-peak value

   f. Effective value

   g. DC level

2. Given $v = 8 \times 10^{-3} \sin 2\pi 4000t$, determine
   a. f

   b. ω

   c. T

   d. Peak value

   e. Peak-to-peak value

   f. Effective value

g. DC level

3. Given V_{rms} = 1.2 V and a frequency of 400 Hz, determine the mathematical expression for the sinusoidal voltage as a function of time.

Part 6. Computer Exercise

PSpice Simulation 1-1

Construct the PSpice circuit shown. Enter the values VAMPL and FREQ for the sinusoidal voltage source V1 given in Problem 3. Set VOFF = 0 V. Run a time (transient) analysis of 5 milliseconds duration. Obtain a Probe plot of the sinusoidal voltage calculated in Problem 3.

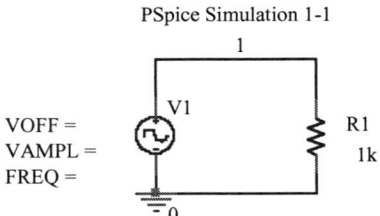

PSpice Simulation 1-1

VOFF =
VAMPL =
FREQ =

Name _____
Date _____
Instructor _____

EXPERIMENT 2

Diode Characteristics

OBJECTIVE

To calculate, compare, draw, and measure the characteristics of a silicon and a germanium diode.

EQUIPMENT REQUIRED

Instruments

DMM

Components

Resistors

(1) 1-kΩ
(1) 1-MΩ

Diodes

(1) Silicon
(1) Germanium

Supplies

DC power supply

Miscellaneous

Demonstration: 1 heat gun

13

EQUIPMENT ISSUED

Item	Laboratory serial no.
DMM	
DC power supply	

RÉSUMÉ OF THEORY

Most modern-day digital multimeters can be used to determine the operating condition of a diode. They have a scale denoted by a diode symbol that will indicate the condition of a diode in the forward- and reverse-bias regions. If connected to establish a forward-bias condition, the meter will display the forward voltage across the diode at a current level typically in the neighborhood of 2 mA. If connected to establish a reverse-bias condition, an "OL" should appear on the display to support the open-circuit approximation frequently applied to this region. If the meter does not have the diode-checking capability, the condition of the diode can also be checked by obtaining some measure of the resistance level in the forward- and reverse-bias regions. Both techniques for checking a diode will be introduced in the first part of the experiment.

The current–volt characteristics of a silicon or germanium diode have the general shape shown in Fig. 2.1. Note the change in scale for both the vertical and horizontal axes. In the reverse-biased region the reverse saturation currents are fairly constant from 0 V to the Zener potential. In the forward-bias region the current increases quite rapidly with increasing diode voltage. Note that the curves are rising almost vertically at a forward-biased voltage of less than 1 V. The forward-biased diode current will be limited solely by the network in which the diode is connected or by the maximum current or power rating of the diode.

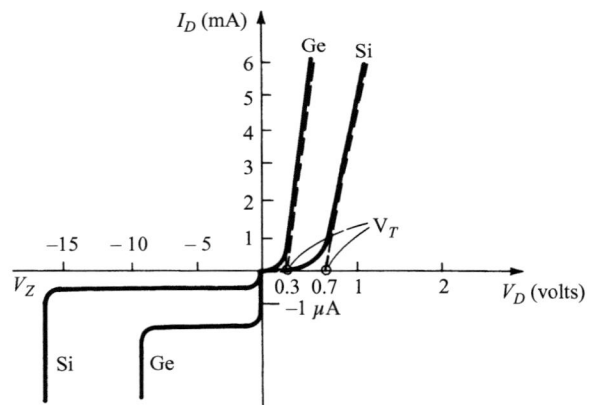

The "firing potential" or threshold voltage is determined by extending a straight line (dashed lines of Fig. 2.1) tangent to the curves until it hits the horizontal axis. The intersection with the V_D axis will determine the threshold voltage V_T at which the current begins to rise rapidly.

Figure 2-1 Silicon and germanium diode characteristics.

The *DC* or *static resistance* of a diode at any point on the characteristics is determined by the ratio of the diode voltage at that point, divided by the diode current. That is,

$$R_{\text{DC}} = \frac{V_D}{I_D} \quad \text{ohms} \tag{2.1}$$

The *AC resistance* at a particular diode current or voltage can be determined using a tangent line drawn as shown in Fig. 2.2. The resulting voltage (ΔV) and current (ΔI) deviations can then be measured and the following equation applied.

$$r_d = \frac{\Delta V}{\Delta I} \quad \text{ohms} \tag{2.2}$$

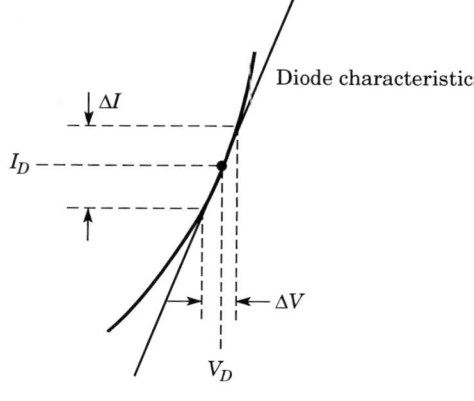

Figure 2-2

The application of differential calculus shows that the AC resistance of a diode in the vertical-rise section of the characteristics is given by

$$r_d = \frac{26 \text{ mV}}{I_D} \quad \text{ohms} \tag{2.3}$$

For levels of current at and below the knee of the curve, the AC resistance of a silicon diode is better approximated by

$$r_d = 2\left(\frac{26 \text{ mV}}{I_D}\right) \quad \text{ohms} \tag{2.4}$$

PROCEDURE

Part 1. Diode Test

Diode Testing Scale

The diode-testing scale of a DMM can be used to determine the operating condition of a diode. With one polarity, the DMM should provide the "firing potential" of the diode, while the reverse connection should result in an "OL" response to support the open-circuit approximation.

Using the connections shown in Fig. 2.3, the constant-current source of about 2 mA internal to the meter will forward bias the junction, and a voltage of about 0.7 V (700 mV) will be obtained for silicon and 0.3 V (300 mV) for germanium. If the leads are reversed, an OL indication will be obtained.

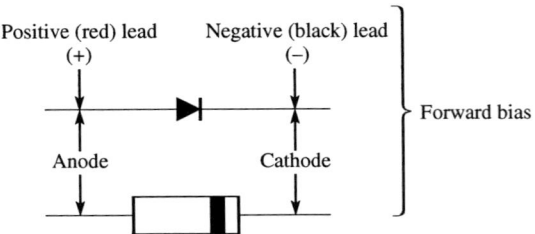

Figure 2-3 Diode testing.

If a low reading (less than 1 V) is obtained in both directions, the junction is shorted internally. If an OL indication is obtained in both directions, the junction is open.

Perform the tests of Table 2.1 for the silicon and germanium diodes.

TABLE 2.1

Test	Si	Ge
Forward		
Reverse		

Based on the results of Table 2.1, are both diodes in good condition?

Resistance Scales

As indicated in the Résumé of Theory section of this experiment, the condition of a diode can also be checked using the resistance scales of a volt-ohm-meter (VOM) or digital meter. Using the appropriate scales of the VOM or DMM, determine the resistance levels of the forward- and reverse-bias regions of the Si and Ge diodes. Enter the results in Table 2.2.

TABLE 2.2

Test	Si	Ge	Meter
Forward			VOM
Reverse			DMM

Although the firing potential is not revealed using the resistance scales, a "good" diode will result in a lower resistance level in the forward bias state and a much higher resistance level when reverse-biased.

Based on the results of Table 2.2, are both diodes in good condition?

Part 2. Forward-bias Diode Characteristics

In this part of the experiment we will obtain sufficient data to plot the forward-bias characteristics of the silicon and germanium diodes on Fig. 2.5.

    a. Construct the network of Fig. 2.4 with the supply (E) set at 0 V. Record the measured value of the resistor.

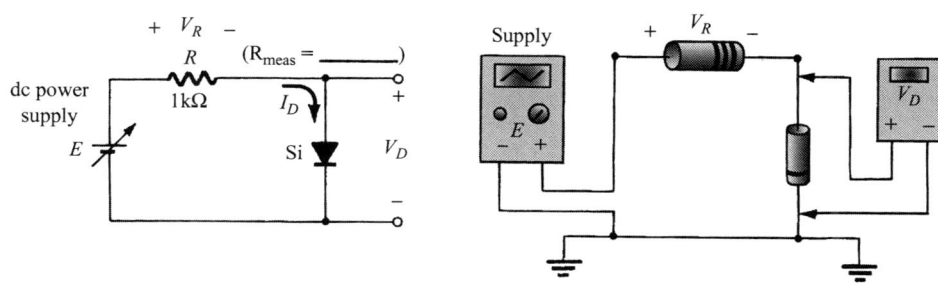

Figure 2-4

    b. Increase the supply voltage E until V_R (not E) reads 0.1 V. Then measure V_D and insert its voltage in Table 2.3. Calculate the value of the corresponding current I_D using the equation shown in Table 2.3.

TABLE 2.3
V_D versus I_D for the silicon diode

V_R (V)	0.1	0.2	0.3	0.4	0.5	0.6	0.7	0.8
V_D (V)								
$I_D = \dfrac{V_R}{R_{meas}}$ (mA)								

V_R (V)	0.9	1	2	3	4	5	6	7	8	9	10
V_D (V)											
$I_D = \dfrac{V_R}{R_{meas}}$ (mA)											

    c. Repeat step b for the remaining settings of V_R, using the equation in Table 2.3.

    d. Replace the silicon diode by a germanium diode and complete Table 2.4.

TABLE 2.4
V_D versus I_D for the germanium diode

V_R (V)	0.1	0.2	0.3	0.4	0.5	0.6	0.7	0.8
V_D (V)								
$I_D = \dfrac{V_R}{R_{meas}}$ (mA)								

V_R (V)	0.9	1	2	3	4	5	6	7	8	9	10
V_D (V)											
$I_D = \dfrac{V_R}{R_{meas}}$ (mA)											

    e. On Fig. 2.5, plot I_D versus V_D for the silicon and germanium diodes. Complete the curves by extending the lower region of each curve to the intersection of the axis at $I_D = 0$ mA and $V_D = 0$ V. Label each curve and clearly indicate data points. Be neat!

f. How do the two curves differ? What are their similarities?

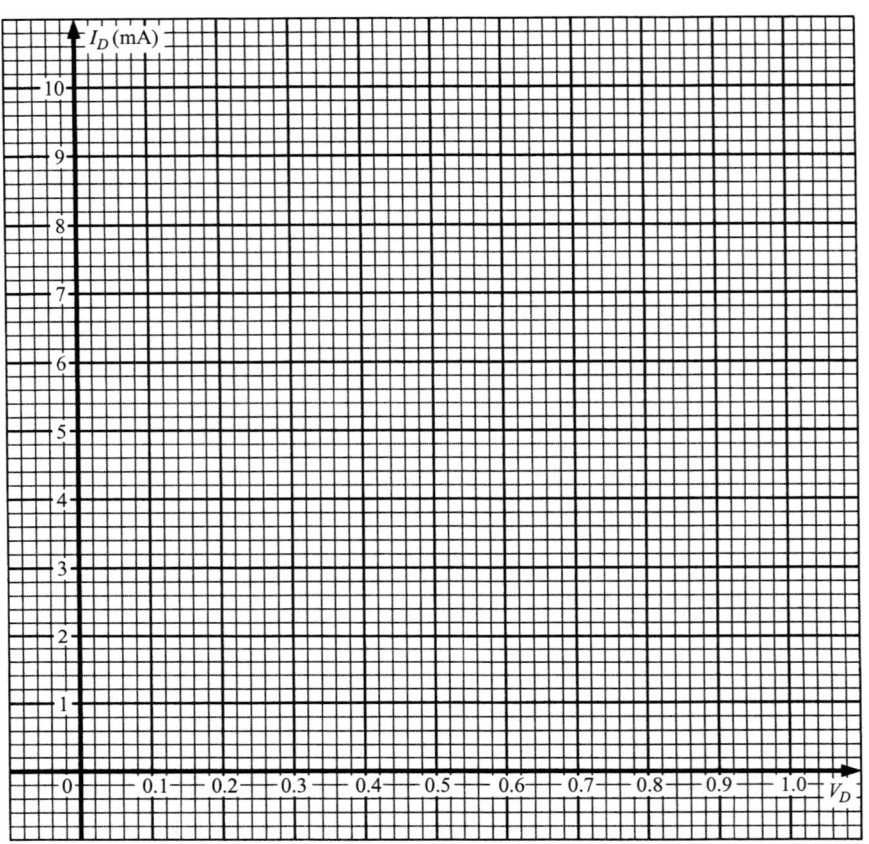

Figure 2-5

Part 3. Reverse Bias

a. In Fig. 2.6 a reverse-bias condition has been established. Since the reverse saturation current will be relatively small, a large resistance of 1 MΩ is required if the voltage across R is to be of measurable amplitude. Construct the circuit of Fig. 2.6 and record the measured value of R on the diagram.

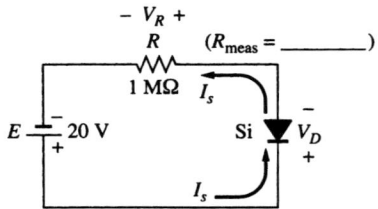

Figure 2-6

b. Measure the voltage V_R. Calculate the reverse saturation current from $I_s = V_R/(R_{meas} || R_m)$. The internal resistance (R_m) of the DMM is included because of the large magnitude of the resistance R. Your instructor will provide the internal resistance of the DMM for your calculations. If unavailable, use a typical value of 10 MΩ.

$R_m =$ _____
V_R (measured) = _____
I_s (calculated) = _____

c. Repeat Part 3(**b**) for the germanium diode.

V_R (measured) = _____
I_s (calculated) = _____

d. How do the resulting levels of I_s for silicon and germanium compare?

e. Determine the DC resistance levels for the silicon and germanium diodes using the equation

$$R_{DC} = \frac{V_D}{I_D} = \frac{V_D}{I_s} = \frac{E - V_R}{I_s}$$

R_{DC} (calculated) (Si) = _____
R_{DC} (calculated) (Ge) = _____

Are the resistance levels sufficiently high to be considered open-circuit equivalents if appearing in series with resistors in the low kilohm range?

Part 4. DC Resistance

a. Using the Si curve of Fig. 2.5, determine the diode voltage at the diode current levels indicated in Table 2.5. Then determine the DC resistance at each current level. Show all calculations.

TABLE 2.5

I_D(mA)	V_D	R_{DC}
0.2		
1		
5		
10		

b. Repeat Part 4(a) for germanium and complete Table 2.6 (Table 2.6 is the same as Table 2.5).

TABLE 2.6

I_D(mA)	V_D	R_{DC}
0.2		
1		
5		
10		

c. Does the resistance (for Si and Ge) change as the diode current increases and we move up the vertical-rise section of the characteristics?

Part 5. AC Resistance

a. Using the equation $r_d = \Delta V/\Delta I$ (Eq. 2.2), determine the AC resistance of the silicon diode at $I_D = 9$ mA using the curve of Fig. 2.5. Show all work.

r_d (calculated) = _____

b. Determine the AC resistance at $I_D = 9$ mA using the equation $r_d = 26$ mV/I_D (mA) for the silicon diode. Show all work.

r_d (calculated) = _____

How do the results of Parts 5(**a**) and 5(**b**) compare?

c. Repeat Part 5(**a**) for $I_D = 2$ mA for the silicon diode.

r_d (calculated) = _____

d. Repeat Part 5(**b**) for $I_D = 2$ mA for the silicon diode. Use Eq. 2.4.

r_d (calculated) = _____

How do the results of Parts 5(**c**) and 5(**d**) compare?

Part 6. Firing Potential

Graphically determine the firing potential (threshold voltage) of each diode from its characteristics as defined in the Résumé of Theory. Show the straight-line approximations on Fig. 2.5.

V_T (silicon) = _____

V_T (germanium) = _____

Part 7. Temperature Effects (Demonstration)

Reconstruct the circuit of Fig. 2.4 using the silicon diode. Establish a current of about 1 mA by setting V_R to 1 V.

a. Place the DMM across the diode. Select the suitable volt scale. Note the reading as the instructor heats the diode with the heat gun. Record the effect on V_D of heating the diode.

b. Let the diode cool down and then measure the voltage across the resistor R. Note the effect on V_R of heating the diode. Since $I_D = V_R/R$ what effect on the diode current of the network results from heating the diode?

c. Since $R_{diode} = V_D/I_D$, what is the effect of increasing temperature on the resistance of the diode?

d. Does a semiconductor diode have a positive or negative temperature coefficient? Explain.

Part 8. Questions

1. Compare the characteristics of silicon and germanium in the forward- and reverse-bias regions. In particular, which diode is closer to the short-circuit approximation in the forward-bias region and which is closer to the open-circuit approximation in the reverse-bias region? How are they similar and what are their most noticeable differences?

2. Research the effect of heat on the terminal resistance of semiconductor materials and briefly review why the terminal resistance will decrease with the application of heat.

Part 9. Computer Exercise

PSpice Simulation 2-1

For the circuit shown, the voltage source V1 changes its voltage from 1 volt to 10 volts in steps of .2 volts. For this operating condition, perform the following steps and answer all questions:

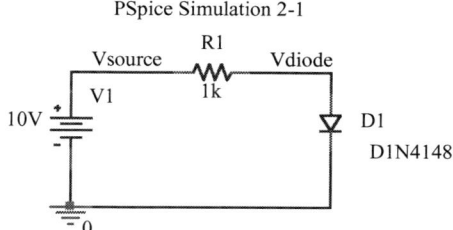

PSpice Simulation 2-1

1. Obtain a plot of the diode current versus the diode voltage Vdiode.

2. Determine the DC resistance at the diode voltages of 600 mV and 700 mV.

3. Compare the two values of the DC resistance at these two voltages.

4. Determine the AC resistance at the diode voltage of 600 mV.

5. Graphically, determine the value of the firing potential.

6. Compare that resistance to the two DC resistances obtained.

7. Based on the data obtained in this problem, is the D1N4148 the model for a germanium or a silicon diode?

8. Perform a temperature analysis at 27°C, 100°C and 200°C. Plot the diode voltage and the diode current as functions of those temperatures. Hint: Run a transient (time) analysis with the auto-range option set for the x-axis.

9. What is the effect of an increase in temperature on both the diode voltage and its current?

10. What are their values at these temperatures?

EXPERIMENT 3

Series and Parallel Diode Configurations

OBJECTIVE

To analyze networks with diodes in a series or parallel configuration and to calculate and measure the circuit voltages of various diode circuits.

EQUIPMENT REQUIRED

Instruments

DMM

Components

Resistors

(1) 1-kΩ
(2) 2.2-kΩ

Diodes

(2) Silicon
(1) Germanium

Supplies

DC power supply

EQUIPMENT ISSUED

Item	Laboratory serial no.
DC power supply	
DMM	

25

RÉSUMÉ OF THEORY

The analysis of circuits with diodes and a DC input requires that the state of the diodes first be determined. For silicon diodes (with a transition voltage or "firing potential" of 0.7 V), the voltage across the diode must be at least 0.7 V with the polarity appearing in Fig. 3.1a for the diode to be in the "on" state. Once the voltage across the diode reaches 0.7 V the diode will turn "on" and have the electrical equivalent of Fig. 3.1b. For $V_D < 0.7$ V or for voltages with the opposite polarity of Fig. 3.1a, the diode can be approximated as an open circuit. For germanium diodes, replace the transition voltage by the germanium value of 0.3 V.

In most networks where the applied DC voltage exceeds the transition voltage of the diodes, the state of the diode can usually be determined simply by mentally replacing the diode by a resistor and determining the direction of current through the resistor. If the direction matches the arrowhead of the diode symbol, the diode is in the "on" state, and if the opposite, it is in the "off" state. Once the state is determined, simply replace the diode by the transition voltage or open circuit and analyze the rest of the network.

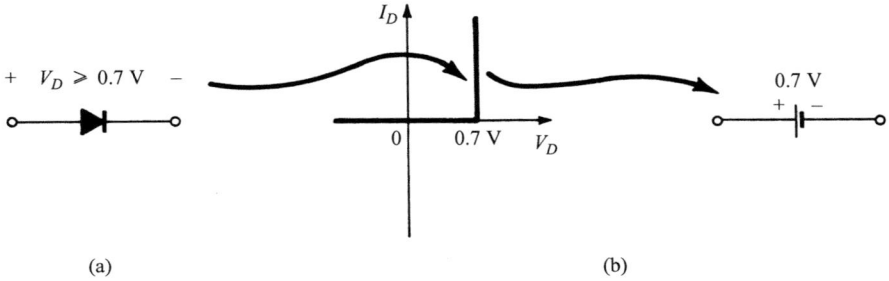

Figure 3-1 Forward-biased silicon diode.

Be alert to the location of the output voltage $V_o = V_R = I_R R$. This is particularly helpful in situations where a diode is in an open-circuit condition and the current is zero. For $I_R = 0$, $V_o = V_R = I_R R = 0(R) = 0$ V. In addition, an open circuit can have a voltage across it, but the current is zero. Further, a short circuit has a zero-volt drop across it, but the current is limited only by the external network or limitations of the diode.

The analysis of logic gates requires that one make an assumption about the state of the diodes, determine the various voltage levels, and then determine whether the results violate any basic laws, such as that a point in a network (such as V_o) can have only one voltage level. It is usually helpful to keep in mind that there must be a forward-bias voltage across a diode equal to the transition voltage to turn it "on." Once V_o is determined and no laws are violated with the diodes in their assumed state, a solution to the configuration can be assumed.

PROCEDURE

Part 1. Threshold Voltage V_T

For both the silicon and the germanium diode, determine the threshold using the diode-checking capability of the DMM or a curve tracer. For this experiment the "firing voltages" obtained will establish the equivalent characteristics for each diode appearing in Fig. 3.2. Record the value of V_T

obtained for each diode in Fig. 3.2. If the diode-checking capability or curve tracer is unavailable, assume $V_T = 0.7$ V for silicon and $V_T = 0.3$ V for germanium.

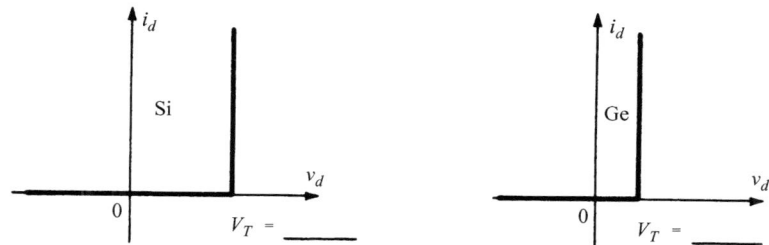

Figure 3-2 Firing voltage for silicon and germanium.

Part 2. Series Configuration

a. Construct the circuit of Fig. 3.3. Record the measured value of R.

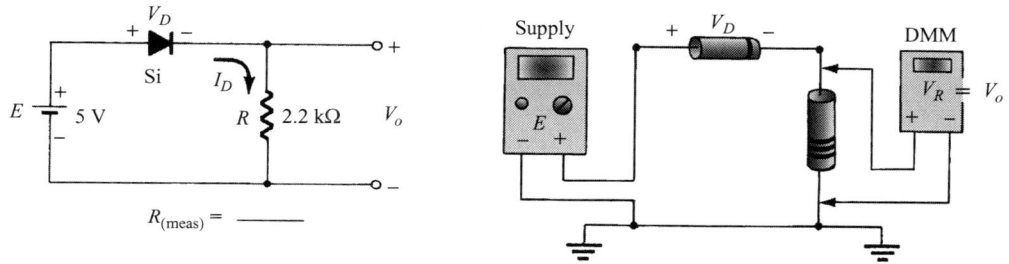

Figure 3-3

b. Using the firing voltages of the silicon and germanium diodes as measured in Part 1 and the measured resistance for R, calculate the theoretical values of V_o and I_D. Insert the level of V_T for V_D.

$V_D =$ _____
V_o (calculated) = _____
I_D (calculated) = _____

c. Measure the voltages V_D and V_o, using the DMM. Calculate the current I_D from measured values. Compare with the results of Part 2(b).

V_D (measured) = _____
V_o (measured) = _____

I_D (from measured) = $\dfrac{V_o}{R}$ = _____

d. Construct the circuit of Fig. 3.4. Record the measured values for each resistor.

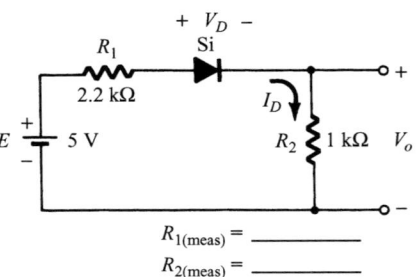

$R_{1(meas)}$ = _____
$R_{2(meas)}$ = _____

Figure 3-4

e. Using the measured values of V_D and V_o from Part 1 and the measured resistance values for R_1 and R_2, calculate the theoretical values of V_o and I_D. Insert the level of V_T for V_D.

V_D = _____
V_o (calculated) = _____
I_D (calculated) = _____

f. Measure the voltages V_D and V_o, using the DMM. Calculate the current I_D from measured values. Compare with the results of step 2(e).

V_D (measured) = _____
V_o (measured) = _____

I_D (from measured) = $\dfrac{V_o}{R_2}$ = _____

g. Reverse the silicon diode in Fig. 3.4 and calculate the theoretical values of V_D, V_o, and I_D.

$V_D =$ _____

V_o (calculated) = _____

I_D (calculated) = _____

h. Measure V_D and V_o for the conditions of Part 2(**g**). Calculate the current I_D from measured values. Compare with the results of Part 2(**g**).

V_D (measured) = _____

V_o (measured) = _____

I_D (from measured) = $\dfrac{V_o}{R_2}$ = _____

i. Construct the network of Fig. 3.5. Record the measured value of R.

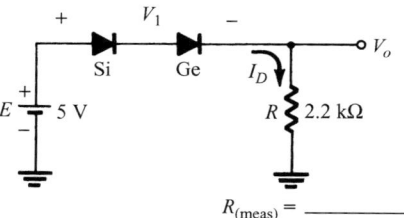

$R_{(meas)} =$ _____

Figure 3-5

j. Using the firing voltages of the silicon and germanium diodes as measured in Part 1, calculate the theoretical values of V_1 (across both diodes), V_o, and I_D.

V_1 (calculated) = _____

V_o (calculated) = _____

I_D (calculated) = _____

k. Measure V_1 and V_o, and compare to the results of Part 2(**j**). Calculate the current I_D from measured values and compare to the level of Part 2(**j**).

V_1 (measured) = _____

V_o (measured) = _____

I_D (from measured) = $\dfrac{V_o}{R}$ = _____

Part 3. Parallel Configuration

a. Construct the network of Fig. 3.6. Record the measured value of R.

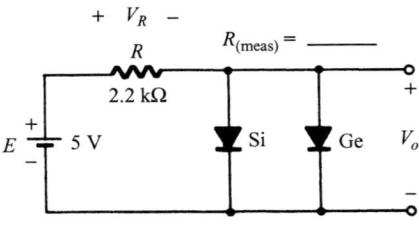

Figure 3-6

b. Using the firing voltages of the silicon and germanium diodes as measured in Part 1, calculate the theoretical values of V_o and V_R.

V_o (calculated) = _____

V_R (calculated) = _____

c. Measure V_o and V_R and compare with the results of Part 3(b).

V_o (measured) = _____

V_R (measured) = _____

d. Construct the network of Fig. 3.7. Record the measured value of each resistor.

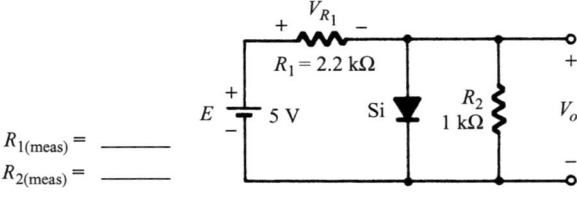

Figure 3-7

e. Using the firing voltages of the silicon diode as measured in Part 1, calculate the theoretical values of V_o, V_{R_1}, and I_D.

V_o (calculated) = _____
V_{R_1} (calculated) = _____
I_D (calculated) = _____

f. Measure V_o and V_{R_1}. Using the measured values of V_o and V_{R_1}, calculate I_{R_2} and I_{R_1} and determine I_D. Compare to the results of Part 3(e).

V_o (measured) = _____
V_{R_1} (measured) = _____
I_D (from measured) = _____

g. Construct the network of Fig. 3.8. Record the measured value of the resistor.

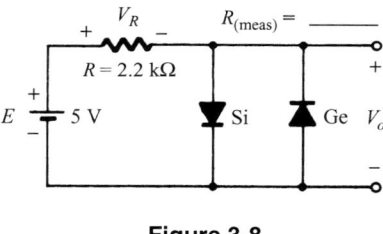

$R_{(meas)}$ = _____

Figure 3-8

h. Using the firing voltages of the silicon and germanium diodes as measured in Part 1, calculate the theoretical values of V_o and V_R.

V_o (calculated) = _____
V_R (calculated) = _____

i. Measure V_o and V_R and compare with the results of Part 3(h).

V_o (measured) = _____
V_R (measured) = _____

Part 4. Positive Logic AND Gate

a. Construct the network of Fig. 3.9. Record the measured value of the resistor.

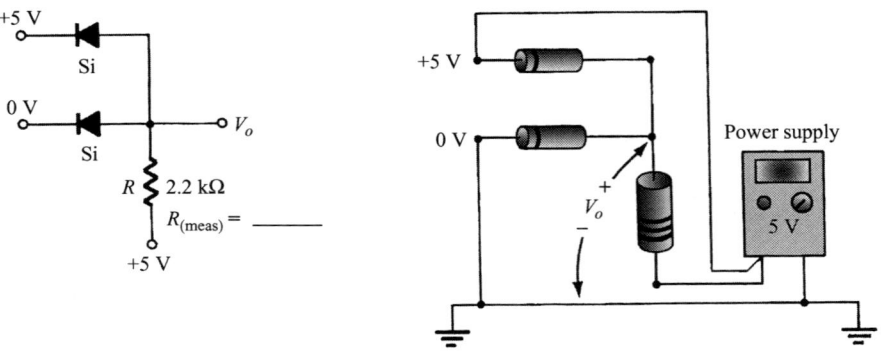

Figure 3-9

b. Using the V_T for both diodes as measured in Part 1, calculate the theoretical value of V_o.

V_o (calculated) = _____

c. Measure V_o and compare to Part 4(**b**).

V_o (measured) = _____

d. Apply 5 V to each input terminal of Fig. 3.9 and calculate the theoretical value of V_o.

V_o (calculated) = _____

e. Measure V_o and compare to the results of Part 4(**d**).

V_o (measured) = _____

f. Set both inputs to zero in Fig. 3.9 (by connecting both inputs to circuit ground) and calculate the theoretical value of V_o.

V_o (calculated) = _____

g. Measure V_o and compare to the results of Part 4(**f**).

V_o (measured) = _____

Part 5. Bridge Configuration

a. Construct the network of Fig. 3.10. Record the measured value of each resistor.

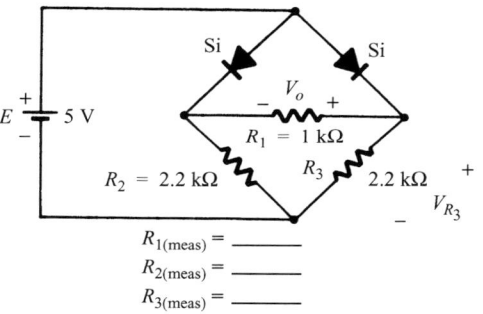

$R_{1(meas)}$ = _____
$R_{2(meas)}$ = _____
$R_{3(meas)}$ = _____

Figure 3-10

b. Using the V_T for both diodes as measured in Part 1, calculate the theoretical value of V_o and V_{R_3}.

V_o (calculated) = _____
V_{R_3} (calculated) = _____

c. Measure V_o and V_{R_3} and compare to the results of Part 5(**b**). Use a low voltage scale when measuring V_o.

V_o (measured) = _____
V_{R_3} (measured) = _____

Part 6. Practical Exercise

a. If the diode in the top right branch of Fig. 3.10 was damaged, creating an internal open-circuit, calculate the resulting levels of V_o and V_{R_3}.

V_o (calculated) = _____
V_{R_3} (calculated) = _____

b. Remove the top right diode from Fig. 3.10 and measure V_o and V_{R_3}. Compare the results with those predicted in Part 6(**a**).

V_o (measured) = _____
V_{R_3} (measured) = _____

Part 7. Computer Exercises

PSpice Simulation 3-1

Analyze the network of Fig. 3.4 using PSpice. Compare the results with those obtained in Part 2(**f**).

Computer V_o = _____ V_o [Part 2(**f**)] = _____
Computer I_D = _____ I_D [Part 2(**f**)] = _____

PSpice Simulation 3-2

The circuit shown in Fig. 3.11 is a PSpice simulation of Figure 3.9. Run a bias point analysis. This will print out the DC voltages of this circuit on the schematic page.

Exp. 3 / Part 7. Computer Exercises

1. Compare Vout obtained by this analysis with both Vo (calculated) and Vo (measured) of Parts 4(b) and 4(c).

2. Compare the value of the voltage Vout with that of the firing potential obtained graphically in PSpice Simulation 2-1, part 5.

PSpice Simulation 3-2

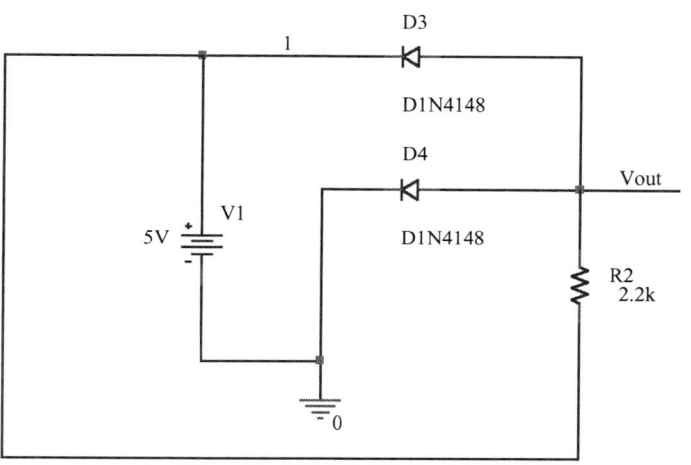

Figure 3-11

Name
Date
Instructor

EXPERIMENT 4

Half-Wave and Full-Wave Rectification

OBJECTIVE

To calculate, draw, and measure the DC output voltages of half-wave and full-wave rectifier circuits.

EQUIPMENT REQUIRED

Instruments

Oscilloscope
DMM

Components

Resistors

(2) 2.2-kΩ
(1) 3.3-kΩ

Diodes

(4) Silicon

Supplies

Function generator

Miscellaneous

12.6-V center-tapped transformer with fused line cord

Exp. 4 / Half-Wave and Full-Wave Rectification

EQUIPMENT ISSUED

Item	Laboratory serial no.
Oscilloscope	
DMM	
Function generator	

RÉSUMÉ OF THEORY

The primary function of half-wave and full-wave rectification systems is to establish a DC level from a sinusoidal input signal that has zero average (DC) level.

The half-wave voltage signal of Fig. 4.1, normally established by a network with a single diode, has an average or equivalent DC voltage level equal to 31.8% of the peak voltage V_m.

That is,

$$\boxed{V_{dc} = 0.318 V_{peak} \text{ volts}} \quad (4.1)$$
half-wave

The full-wave rectified signal of Fig. 4.2 has twice the average or DC level of the half-wave signal, or 63.6% of the peak value V_m.

That is,

$$\boxed{V_{dc} = 0.636 V_{peak} \text{ volts}} \quad (4.2)$$
full-wave

For large sinusoidal inputs ($V_m >> V_T$) the forward-biased transition voltage V_T of a diode can be ignored. However, for situations when the peak value of the sinusoidal signal is not that much greater than V_T, V_T can have a noticeable effect on V_{DC}.

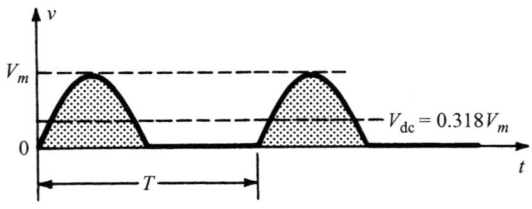

Figure 4-1

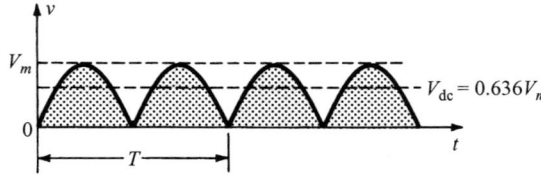

Figure 4-2 Full-wave rectified signal.

In rectification systems the peak inverse voltage (PIV) must be considered carefully. The PIV voltage is the maximum reverse-bias voltage that a diode can handle before entering the Zener breakdown region. For typical single-diode half-wave rectification systems, the required PIV level is equal to the peak value of the applied sinusoidal signal. For the four-diode full-wave bridge rectification system, the required PIV level is again the peak value, but for a two-diode center-tapped configuration, it is twice the peak value of the applied signal.

Exp. 4 / Procedure

PROCEDURE

Part 1. Threshold Voltage

Choose one of the four silicon diodes and determine the threshold voltage, V_T, using the diode-checking capability of the DMM or a curve tracer.

$V_T =$ _____

Part 2. Half-Wave Rectification

a. Construct the circuit of Fig. 4.3 using the chosen diode of Part 1. Record the measured value of the resistance R. Set the function generator to a 1000-Hz 8-V_{p-p} sinusoidal voltage using the oscilloscope.

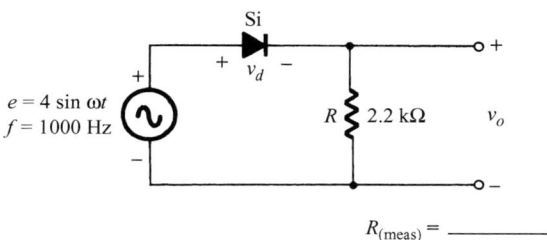

$R_{(meas)} =$ _____

Figure 4-3 Half-wave rectifier.

b. The sinusoidal input (e) of Fig. 4.3 has been plotted on the screen of Fig. 4.4. Determine the chosen vertical and horizontal sensitivities. Note that the horizontal axis is the 0 V line.

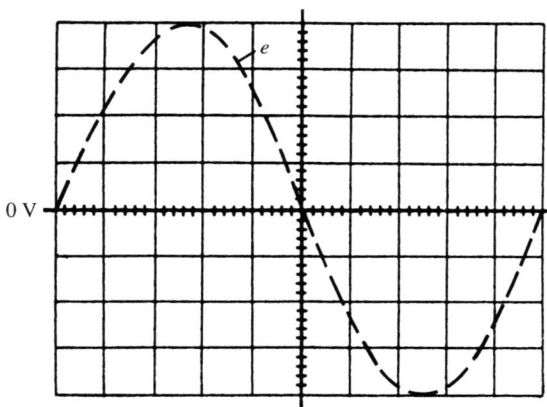

Figure 4-4

Vertical sensitivity = _____

Horizontal sensitivity = _____

c. Using the threshold voltage V_T of Part 1, determine the theoretical output voltage v_o for the circuit of Fig. 4.3 and sketch the waveform on Fig. 4.4 for one full cycle using the same sensitivities employed in Part 2(**b**). Indicate the maximum and minimum values on the output waveform.

d. Using the oscilloscope with the AC-GND-DC coupling switch in the DC position, obtain the voltage v_o and sketch the waveform on Fig. 4.5. Before viewing v_o be sure to set the $v_o = 0$ V line using the GND position of the coupling switch. Use the same sensitivities as in Part 2(**b**).

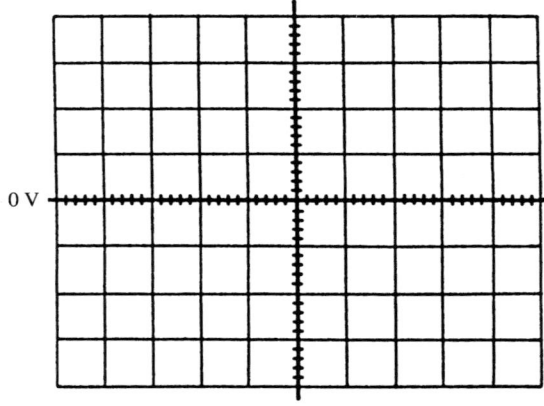

Figure 4-5

How do the results of Parts 2(**c**) and 2(**d**) compare?

e. Calculate the DC level of the half-wave rectified signal of Part 2(**d**) using Eq. 4.1.

V_{DC} (calculated) = _____

f. Measure the DC level of v_o using the DC scale of the DMM and find the percent difference between the measured value and the calculated value of Part 2(**e**) using the following equation:

$$\% \text{ Difference} = \left| \frac{V_{DC\,(calc)} - V_{DC\,(meas)}}{V_{DC\,(calc)}} \right| \times 100\%$$

V_{DC} (measured) = _____
(% Difference) = _____

Exp. 4 / Procedure

g. Switch the AC-GND-DC coupling switch to the AC position. What is the effect on the output signal v_o? Does it appear that the area under the curve above the zero axis equals the area under the curve below the zero axis? Discuss the effect of the AC position on waveforms that have an average value over one full cycle.

h. Reverse the diode of Fig. 4.3 and sketch the output waveform obtained using the oscilloscope on Fig. 4.6. Be sure the coupling switch is in the DC position and the $v_o = 0$ V line is preset using the GND position. Include the maximum and minimum voltage levels on the plot as determined using the chosen vertical sensitivity.

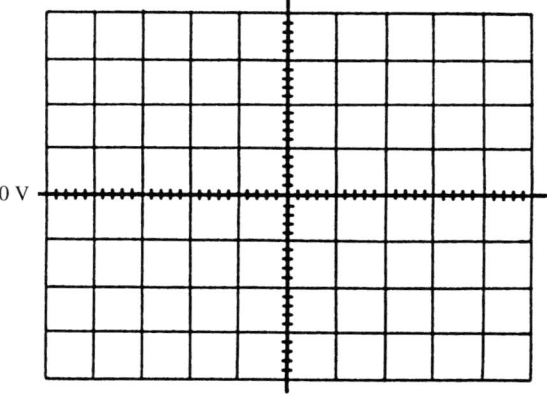

Figure 4-6

i. Calculate and measure the DC level of the resulting waveform of Fig. 4.6. Insert the proper sign for the polarity of V_{DC} as defined by Fig. 4.3 using Eq. 4.1.

V_{DC} (calculated) = _____
V_{DC} (measured) = _____

Part 3. Half-Wave Rectification (continued)

a. Construct the network of Fig. 4.7. Record the measured value of the resistor R.

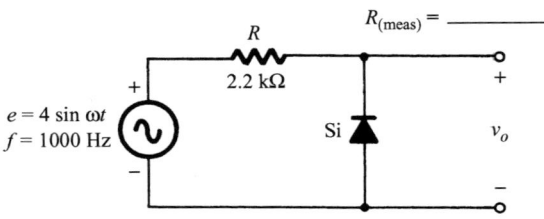

Figure 4-7

b. Using the threshold voltage of Part 1, determine the theoretical output voltage v_o for Fig. 4.7 and sketch the waveform on Fig. 4.8 for one full cycle using the same sensitivities employed in Part 2(**b**). Indicate the maximum and minimum values on the output waveform.

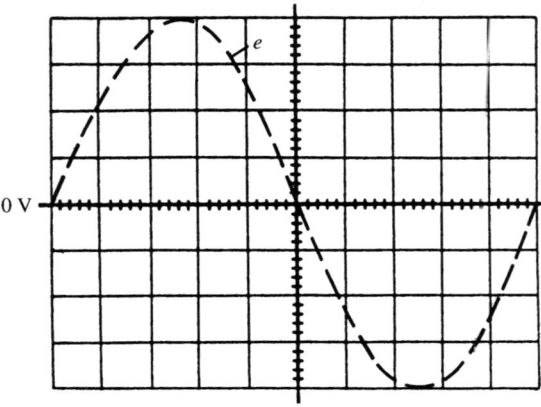

Figure 4-8

c. Using the oscilloscope with the coupling switch in the DC position, obtain the voltage v_o and sketch the waveform on Fig. 4.9. Before viewing v_o be sure to set the $v_o = 0$ V line using the GND position of the coupling switch. Use the same sensitivities as in Part 3(**b**).

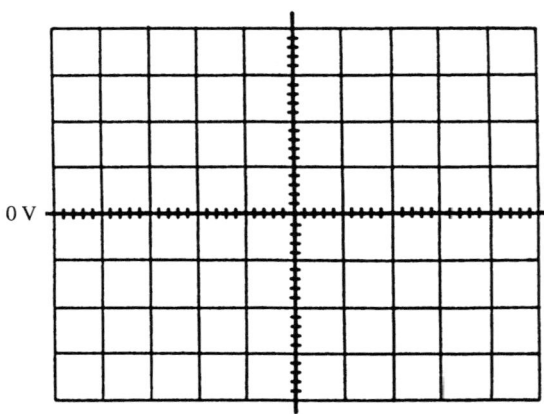

Figure 4-9

How do the results of Parts 3(**b**) and 3(**c**) compare?

d. What is the most noticeable difference between the waveform of Fig. 4.9 and that obtained in Part 2(**h**)? Why did the difference occur?

e. Calculate the DC level of the waveform of Fig. 4.9 using the following equation:

$$V_{DC} = \frac{\text{Total Area}}{2\pi} \cong \frac{2V_m - (V_T)\pi}{2\pi} = 0.318V_m - V_T/2 \text{ volts}$$

V_{DC} (calculated) = _____

f. Measure the output DC voltage with the DC scale of the DMM and calculate the percent difference using the same equation appearing in Part 2(**f**).

V_{DC} (measured) = _____
(% Difference) = _____

Part 4. Half-Wave Rectification (continued)

a. Construct the network of Fig. 4.10. Record the measured value of each resistor.

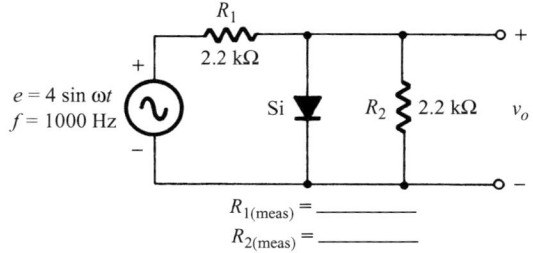

$R_{1(\text{meas})}$ = _____
$R_{2(\text{meas})}$ = _____

Figure 4-10

b. Using the measured resistor values and V_T from Part 1, forecast the appearance of the output waveform v_o and sketch the result on Fig. 4.11. Use the same sensitivities employed in Part 2(**b**) and insert the maximum and minimum values of the waveform.

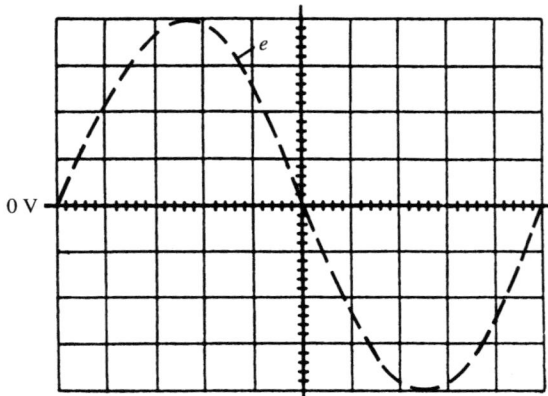

Figure 4-11

c. Using the oscilloscope with the coupling switch in the DC position, obtain the waveform for v_o and record on Fig. 4.12. Again, be sure to preset the $v_o = 0$ V line using the GND position of the coupling switch before viewing the waveform. Using the chosen sensitivities, determine the maximum and minimum values and place on the sketch of Fig. 4.12.

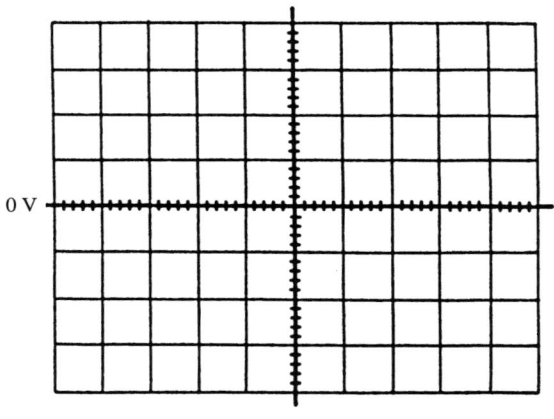

Figure 4-12

Are the waveforms of Figs. 4.11 and 4.12 relatively close in appearance and magnitude?

d. Reverse the direction of the diode and record the resulting waveform on Fig. 4.13 as obtained using the oscilloscope.

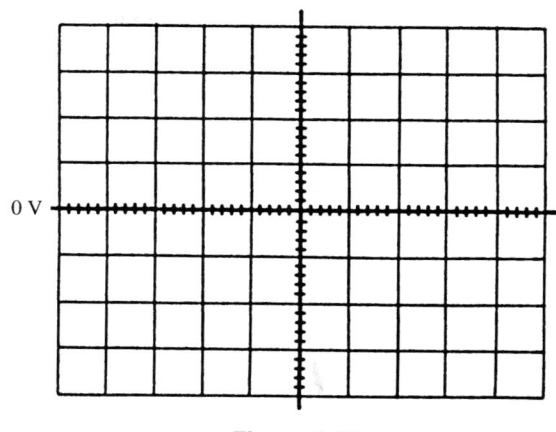

Figure 4-13

Compare the results of Figs. 4.12 and 4.13. What are the major differences and why?

Part 5. Full-Wave Rectification (Bridge Configuration)

a. Construct the full-wave bridge rectifier of Fig. 4.14. Be sure that the diodes are inserted correctly and that the grounding is as shown. If unsure, ask your instructor to check your setup. Record the measured value of the resistor R.

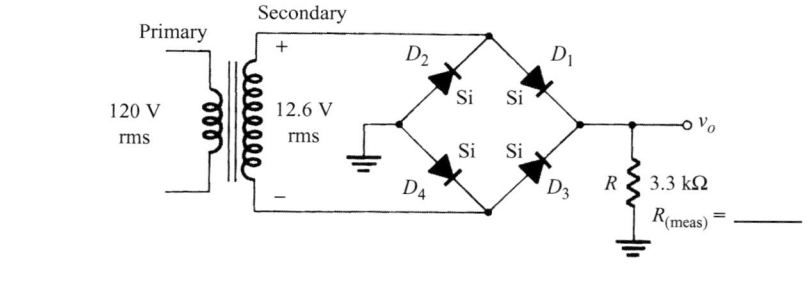

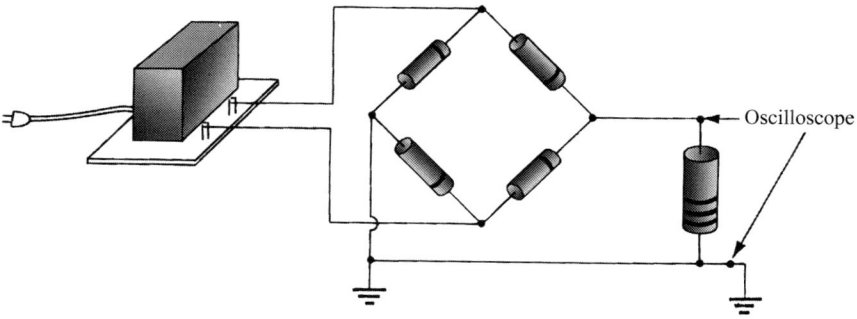

Figure 4-14

In addition, measure the rms voltage at the transformer secondary using the DMM set to AC. Record that rms value below. Does it differ from the rated 12.6 V?

V_{rms} (measured) = _____

b. Calculate the peak value of the secondary voltage using the measured value ($V_{peak} = 1.414 V_{rms}$).

V_{peak} (calculated) = _____

c. Using the V_T of Part 1 for each diode, sketch the expected output waveform v_o on Fig. 4.15. Choose a vertical and a horizontal sensitivity based on the amplitude of the secondary voltage. Consult your oscilloscope to obtain a list of possible sensitivities. Record your choice for each below.

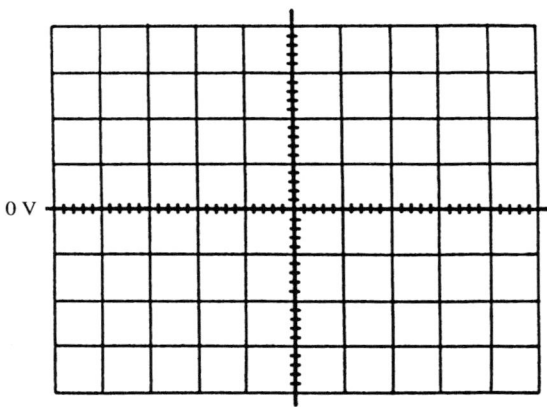

Figure 4-15

Vertical sensitivity = _____
Horizontal sensitivity = _____

d. Using the oscilloscope with the coupling switch in the DC position, obtain the waveform for v_o and record on Fig. 4.16. Use the same sensitivities employed in Part 5(**c**) and be sure to preset the $v_o = 0$ V line using the GND position of the coupling switch. Label the maximum and

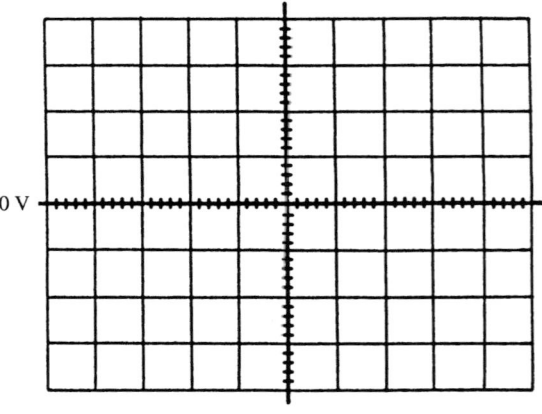

Figure 4-16

minimum values of the waveform using the chosen vertical sensitivity.

How do the waveforms of Parts 5(c) and 5(d) compare?

e. Determine the DC level of the full-wave rectified waveform of Fig. 4.16.

V_{DC} (calculated) = _____

f. Measure the DC level of the output waveform using the DMM and calculate the percent difference between the measured and calculated values.

V_{DC} (measured) = _____
(% Difference) = _____

g. Replace diodes D_3 and D_4 by 2.2 kΩ resistors and forecast the appearance of the output voltage v_o including the effects of V_T for each diode. Sketch the waveform on Fig. 4.17 and label the magnitude of the maximum and minimum values. Record your choice of sensitivities below.

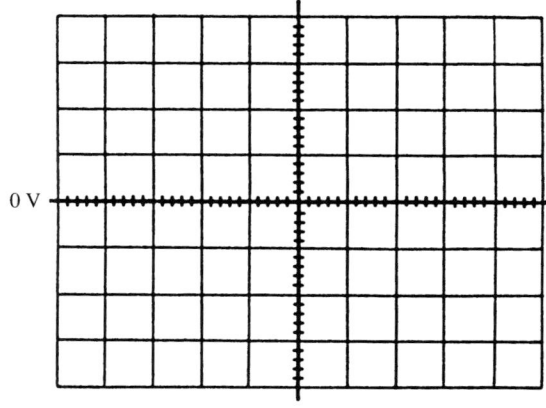

Figure 4-17

Vertical sensitivity = _____
Horizontal sensitivity = _____

h. Using the oscilloscope, obtain the waveform for v_o and reproduce it on Fig. 4.18, indicating the maximum and minimum values. Use the same sensitivities as determined in Part 5(**g**).

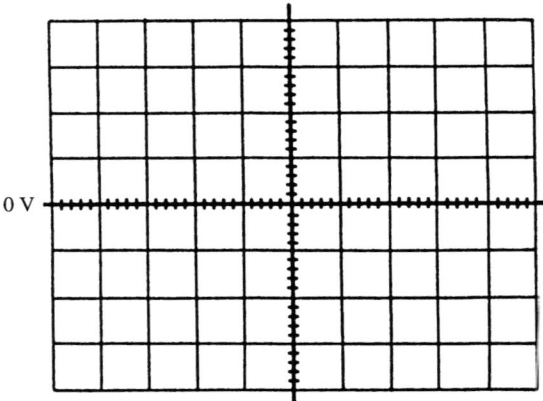

Figure 4-18

How do the waveforms of Figs. 4.17 and 4.18 compare?

i. Calculate the DC level of the waveform of Fig. 4.18.

V_{DC} (calculated) = _____

j. Measure the DC level of the output voltage using the DMM and calculate the percent difference.

V_{DC} (measured) = _____
(% Difference) = _____

k. What was the major effect of replacing the two diodes with resistors?

Part 6. Full-Wave Center-Tapped Configuration

a. Construct the network of Fig. 4.19. Record the measured value of the resistor R.

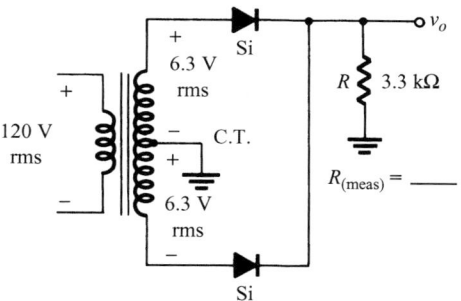

Figure 4-19

Measure the two secondary voltages of the transformer with the DMM set on AC. Record below. Do they differ from the 6.3 V rating?

V_{rms} (measured) = _____
V_{rms} (measured) = _____

Using the average of the two rms readings, calculate the peak value of the overall secondary voltage.

V_{peak} (calculated) = _____

b. Using the V_T of Part 1 for each diode, sketch the expected output waveform v_o on Fig. 4.20. Choose a vertical and a horizontal sensitivity based on the amplitude of the secondary voltage. Record your choice for each below.

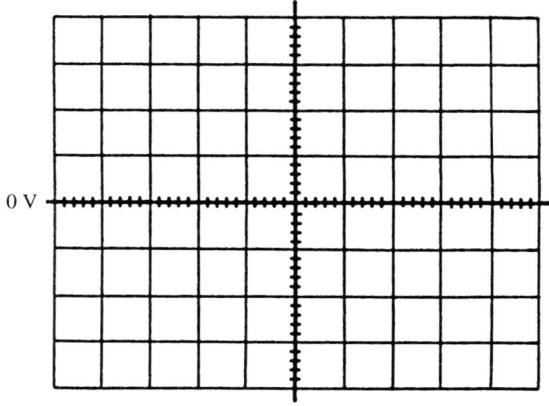

Figure 4-20

Vertical sensitivity = _____
Horizontal sensitivity = _____

c. Using the oscilloscope with the coupling switch in the DC position, obtain the waveform for v_o and record on Fig. 4.21. Use the same sensitivities employed in Part 6(**b**) and be sure to preset the $v_o = 0$ V line using the GND position of the coupling switch. Label the maximum and minimum values of the waveform using the chosen vertical sensitivity.

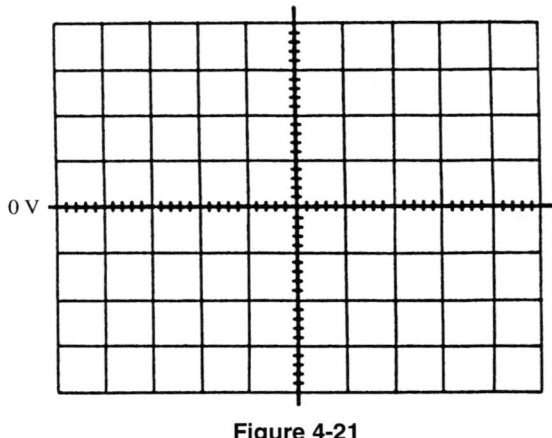

Figure 4-21

How do the waveforms of Figs. 4.20 and 4.21 compare?

d. Determine and compare the calculated and measured values of the DC level associated with v_o.

(calculated) = _____
(measured) = _____

Part 7. Computer Exercises

PSpice Simulation 4-1

Analyze the network of Fig. 4.3 using PSpice Windows. Compare the results with those obtained in Part 2.

PSpice Simulation 4-2

The circuit shown is a PSpice simulation of Fig. 4.19. The peak voltage of 170 volts of the voltage source V1 corresponds to the RMS voltage of the voltage source of Fig. 4.19. The primary windings of transformers TX1 and TX2 have inductances of 100 mH each. The secondary windings of transformers TX1 and TX2 have inductances of 1 mH each. This makes the secondary voltages close to 6 volts rms. Run a Time Domain (transient) analysis of 34 milliseconds duration and answer the following questions (using Probe plots):

PSpice Simulation 4-2

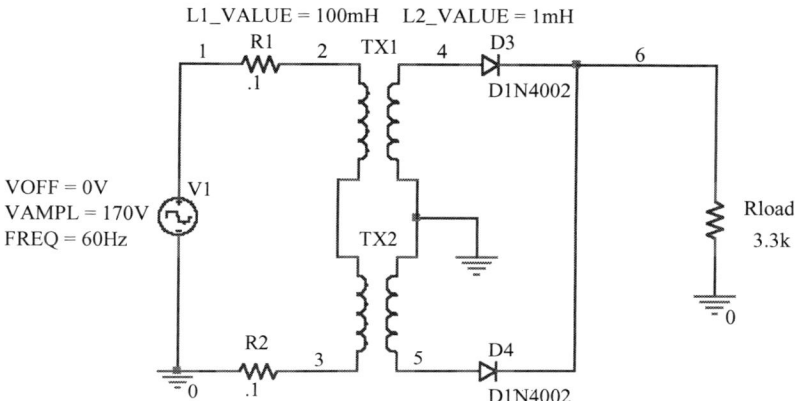

1. What are the amplitudes of the secondary voltages and what is their relative phase shift?

2. Compare the peak-inverse-voltage (PIV) of either diode to the sum of the secondary voltages.

3. Compare the conduction cycle of the two diodes.

4. Obtain a plot of the voltage across Rload. What is its amplitude?

5. Is that voltage consistent with the output voltage of a full-wave rectifier?

6. How do the results of this simulation compare with those of Fig. 4.19?

EXPERIMENT 5

Clipping Circuits

OBJECTIVE

To calculate, draw, and measure the output voltages of series and parallel clipping circuits.

EQUIPMENT REQUIRED

Instruments

Oscilloscope
DMM

Components

Resistors

(1) 2.2-kΩ

Diode

(1) Silicon
(1) Germanium

Supplies

(1) 1.5-V D cell and holder
Function generator

EQUIPMENT ISSUED

Item	Laboratory serial no.
Oscilloscope	
DMM	
Function generator	

Exp. 5 / Clipping Circuits

RÉSUMÉ OF THEORY

The primary function of clippers is to "clip" away a portion of an applied alternating signal. The process is typically performed by a resistor-diode combination. DC batteries are used to provide additional shifts or "cuts" of the applied voltage. The analysis of clippers with square-wave inputs is the easiest to perform since there are only two levels of input voltage. Each level can be treated as a DC input and the output voltage for the corresponding time interval determined. For sinusoidal and triangular inputs, various instantaneous values can be treated as DC levels and the output level determined. Once a sufficient number of plot points for the output voltage v_o has been determined, it can be sketched in total. Once the behavior of clippers is established, the effect of the placement of elements in various positions can be predicted and the analysis completed.

PROCEDURE

Part 1. Threshold Voltage

Determine the threshold voltage for the silicon and germanium diodes using the diode-checking capability of the DMM or a curve tracer. Round off to the hundredths place when recording in the designated space below. If the diode-checking capability or curve tracer is unavailable, assume $V_T = 0.7$ V for the silicon diode and 0.3 V for the germanium diode.

$V_T(\text{Si}) =$ _____

$V_T(\text{Ge}) =$ _____

Part 2. Parallel Clippers

a. Construct the clipping network of Fig. 5.1. Record the measured resistance value and voltage of the D cell. Note that the input is an 8 V_{p-p} square wave at a frequency of 1000 Hz.

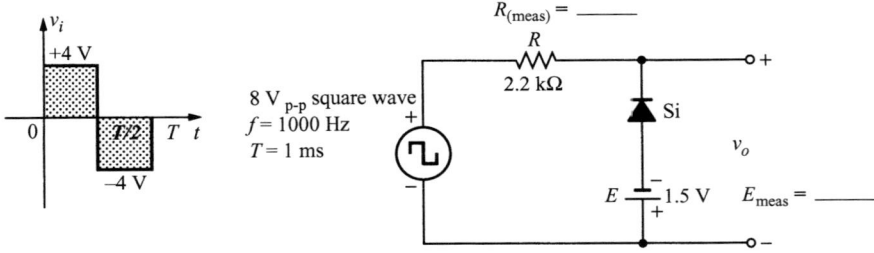

Figure 5-1

b. Using the measured values of R, E, and V_T, calculate the voltage V_o when the applied square wave is +4 V. What is the level of V_o? Show all the steps of your calculations to determine V_o.

V_o (calculated) = _____

c. Repeat Part 2(b) when the applied square wave is –4 V.

V_o (calculated) = _____

d. Using the results of Parts 2(b) and 2(c), sketch the expected waveform for v_o using the horizontal axis of Fig. 5.2 as the $V_o = 0$ V line. Use a vertical sensitivity of 1 V/cm and a horizontal sensitivity of 0.2 ms/cm.

Sketch of V_o from calculated results:

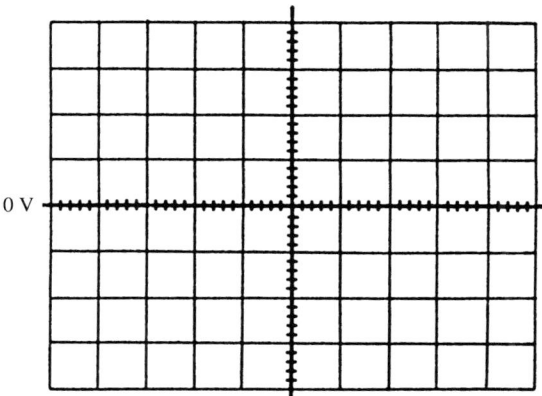

Figure 5-2

e. Using the sensitivities settings provided in Part 2(d), set the input square wave and record v_o on Fig. 5.3 using the oscilloscope. Be sure to preset the $V_o = 0$ V line using the GND position of the coupling switch (and the DC position to view the waveform).

Sketch of V_o from measured results:

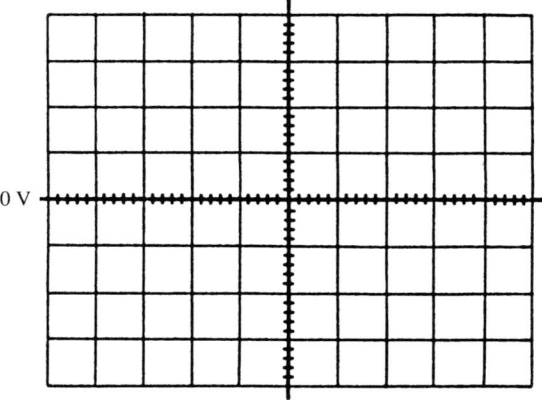

Figure 5-3

How does the waveform of Fig. 5.3 compare with the predicted result of Fig. 5.2?

f. Reverse the battery of Fig. 5.1 and, using the measured values of R, E, and V_T, calculate the level of V_o for the time interval when $V_i = +4$ V.

V_o (calculated) = _____

g. Repeat Part 2(**f**) for the time interval when $V_i = -4$ V.

V_o (calculated) = _____

h. Using the results of Parts 2(**f**) and 2(**g**), sketch the expected waveform for v_o using the horizontal axis of Fig. 5.4 as the $V_o = 0$ V line. Use the same sensitivities provided in Part 2(**d**).

Sketch of V_o from calculated results:

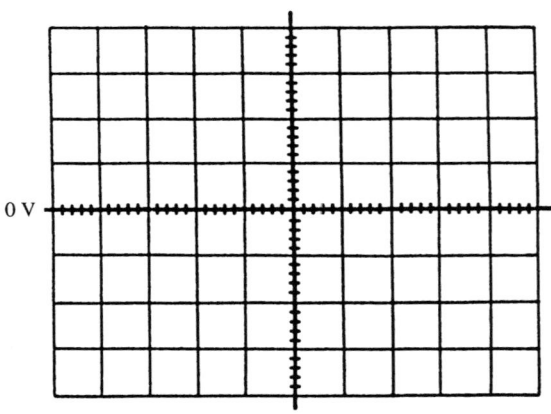

Figure 5-4

i. Set the input square wave and record v_o on Fig. 5.5 using the oscilloscope. Be sure to preset the $V_o = 0$ V line using the GND position of the coupling switch (and the DC position to view the waveform).

How does the waveform of Fig. 5.4 compare with the predicted result of Fig. 5.5?

Sketch of V_o from measured results:

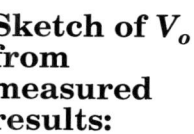

Figure 5-5

Part 3. Parallel Clippers (continued)

a. Construct the network of Fig. 5.6. Record the measured value of the resistance. Note that the input is now a 4 V_{p-p} square wave at $f = 1000$ Hz.

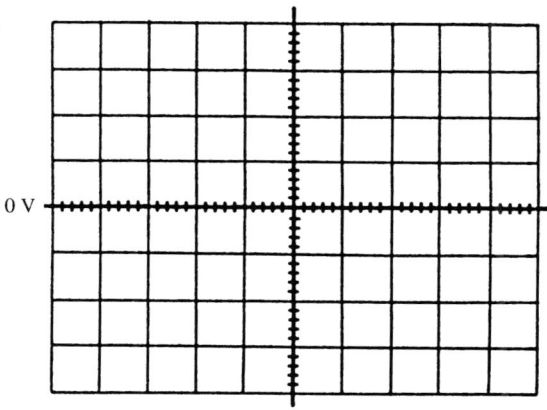

Figure 5-6

b. Using the levels of V_T determined in Part 1, calculate the level of V_o for the time interval when $V_i = +2$ V.

V_o (calculated) = _____

c. Repeat Part 3(b) for the time interval when $V_i = -2$ V.

V_o (calculated) = _____

d. Using the results of Parts 3(b) and 3(c), sketch the expected waveform for v_o using the horizontal axis of Fig. 5.7 as the $V_o = 0$ V line. Insert your chosen vertical and horizontal sensitivities below.

Vertical sensitivity = _____
Horizontal sensitivity = _____

Sketch of V_o from calculated results:

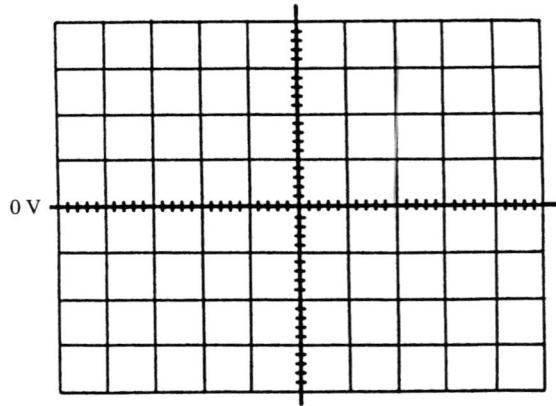

0 V

Figure 5-7

e. Using the sensitivity settings chosen in Part 3(**d**), set the input square wave and record v_o on Fig. 5.8 using the oscilloscope. Be sure to preset the $V_o = 0$ V line using the GND position of the coupling switch (and the DC position to view the waveform).

Sketch of V_o from measured results:

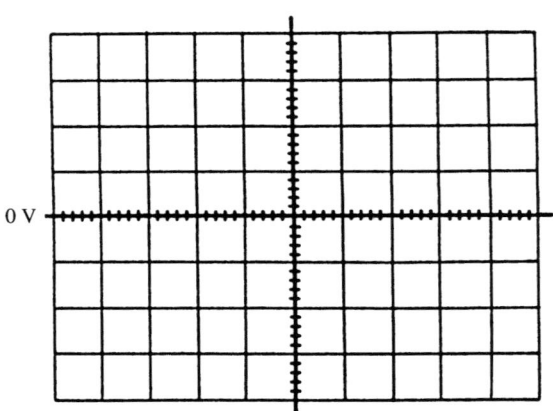

0 V

Figure 5-8

How does the waveform of Fig. 5.8 compare with the predicted result of Fig. 5.7?

Part 4. Parallel Clippers (Sinusoidal Input)

a. Rebuild the circuit of Fig. 5.1 but change the input signal to an 8 V_{p-p} sinusoidal signal with the same frequency (1000 Hz).
b. Using the results of Part 2 and any other analysis technique, sketch the expected output waveform for v_o on Fig. 5.9. In particular, find V_o when the applied signal is at its positive and negative peak and zero volts. Also, list the chosen vertical and horizontal sensitivities below:

Sketch of V_o from calculated results:

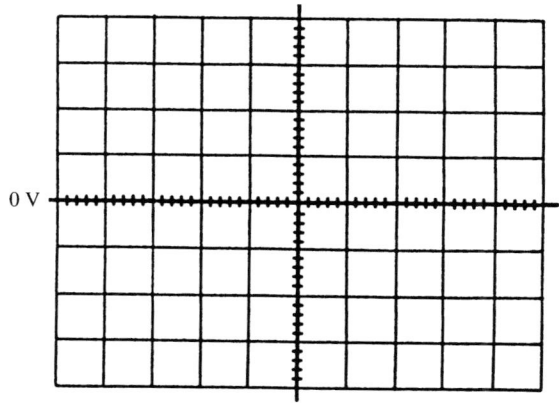

Figure 5-9

V_o (calculated) when $V_i = +4$ V = _____
V_o (calculated) when $V_i = -4$ V = _____
V_o (calculated) when $V_i = 0$ V = _____
Vertical sensitivity = _____
Horizontal sensitivity = _____

c. Using the sensitivity settings chosen in Part 4(**b**), set the input sinusoidal waveform and record v_o on Fig. 5.10 using the oscilloscope. Be sure to preset the $V_o = 0$ V line using the GND position of the coupling switch.

Sketch of V_o from measured results:

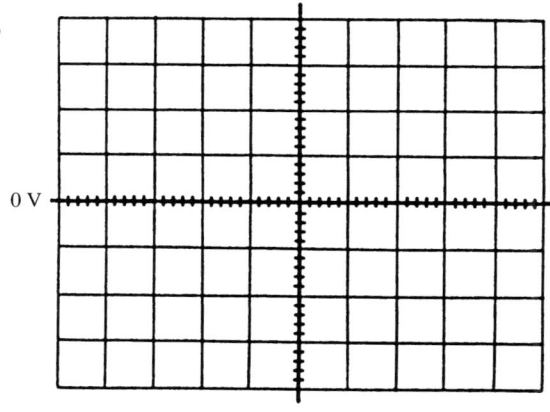

Figure 5-10

How does the waveform of Fig. 5.10 compare with the predicted result of Fig. 5.9?

Part 5. Series Clippers

a. Construct the circuit of Fig. 5.11. Record the measured resistance value and the DC level of the D cell. The applied signal is an 8 V_{p-p} square wave at a frequency of 1000 Hz.

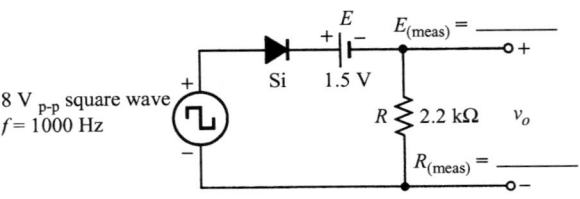

Figure 5-11

b. Using the measured values of R, E, and V_T, calculate the voltage V_o for the time interval when $V_i = +4$ V.

V_o (calculated) = _____

c. Repeat Part 5(**b**) for the time interval when $V_i = -4$ V.

V_o (calculated) = _____

d. Using the results of Parts 5(**b**) and 5(**c**), sketch the expected waveform for v_o using the horizontal axis of Fig. 5.12 as the $V_o = 0$ V line. Insert your chosen vertical and horizontal sensitivity settings below.

Sketch of V_o from calculated results:

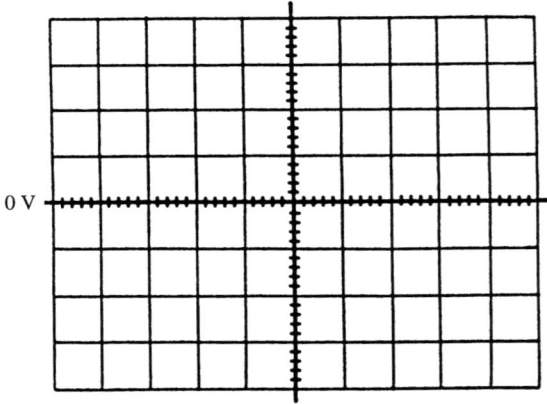

Figure 5-12

Vertical sensitivity = _____
Horizontal sensitivity = _____

e. Using the sensitivities chosen in Part 5(**d**), set the input square wave and record v_o on Fig. 5.13 using the oscilloscope. Be sure to preset the $V_o = 0$ V line using the GND position of the coupling switch (and the DC position to view the waveform).

Sketch of V_o from measured results:

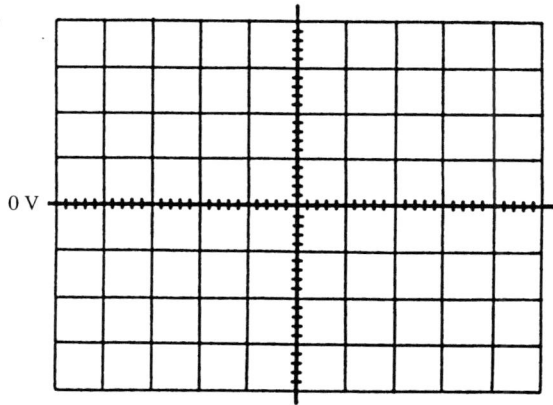

Figure 5-13

How does the waveform of Fig. 5.13 compare with the predicted result of Part 5(**d**)?

f. Reverse the battery of Fig. 5.11 and, using the measured values of R, E, and V_T, calculate the level of V_o for the time interval when $V_i = +4$ V.

V_o (calculated) = _____

g. Repeat Part 5(**f**) for the time interval when $V_i = -4$ V.

V_o (calculated) = _____

h. Using the results of Parts 5(**f**) and 5(**g**), sketch the expected waveform for v_o using the horizontal axis of Fig. 5.14 as the $V_o = 0$ V line. Use the following sensitivities:

Vertical: 2 V/cm
Horizontal: 0.2 ms/cm

Sketch of V_o from calculated results:

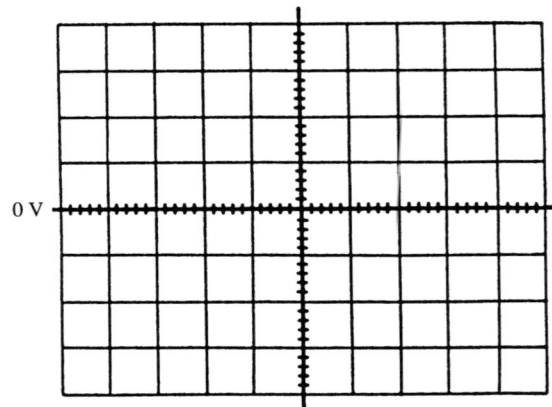

Figure 5-14

i. Using the sensitivities provided in Part 5(**h**), set the input square wave and record v_o on Fig. 5.15 using the oscilloscope. Be sure to preset the $V_o = 0$ V line using the GND position of the coupling switch (and the DC position to view the waveform).

Sketch of V_o from measured results:

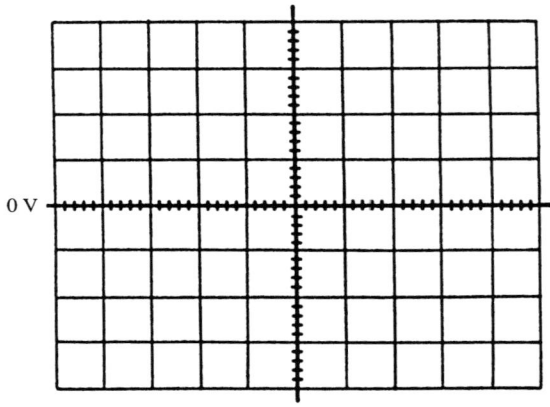

Figure 5-15

How does the waveform of Fig. 5.15 compare with the predicted pattern of Fig. 5.14?

Part 6. Series Clippers (Sinusoidal Input)

a. Rebuild the circuit of Fig. 5.11 but change the input signal to an 8 V_{p-p} sinusoidal signal with the same frequency (1000 Hz).

b. Using the results of Part 5 and any other analysis technique, sketch the expected output waveform for v_o on Fig. 5.16. In particular, find V_o when the applied signal is at its positive and negative peak and zero volts. Use a vertical sensitivity of 1 V/cm and a horizontal sensitivity of 0.2 ms/cm.

Sketch of V_o from calculated results:

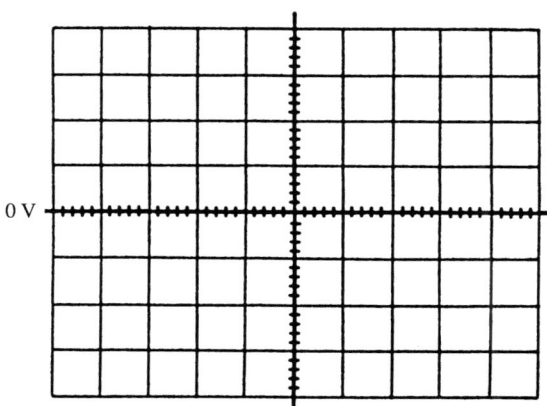

Figure 5-16

V_o (calculated) when $V_i = +4$ V = _____
V_o (calculated) when $V_i = -4$ V = _____
V_o (calculated) when $V_i = 0$ V = _____

c. Using the sensitivities provided in Part 6(**b**), set the input sinusoidal waveform and record v_o on Fig. 5.17 using the oscilloscope. Be sure to preset the $V_o = 0$ V line using the GND position of the coupling switch.

Sketch of V_o from measured results:

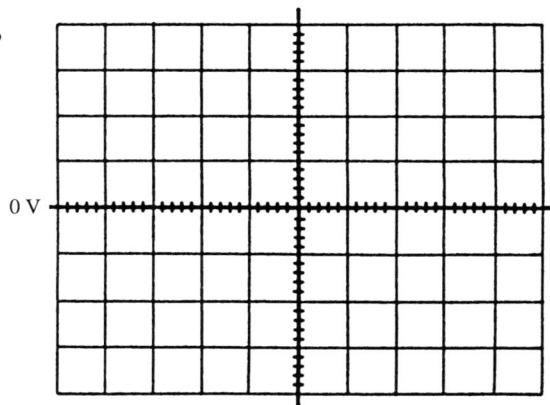

Figure 5-17

How does the waveform of Fig. 5.17 compare with the predicted result of Fig. 5.16?

Part 7. Computer Exercises

PSpice Simulation 5-1

Analyze the network of Fig. 5.1 using PSpice Windows. Compare the computer generated result with that of Part 2.

PSpice Simulation 5-2

The circuit shown is a PSpice simulation of Fig. 5.1. You need to get the Vpulse voltage source from the PSpice program. Its parameters are set as shown. The rise time (TR) and fall time (TF) are chosen so that the pulse transitions with almost vertical slopes. The PSpice model of a diode is not ideal. Thus, neither a short circuit exists in the forward-bias mode, nor does an ideal open circuit exist in the reverse-bias mode. This will effect the value of Vout from its ideal values for the two conducting stages of the diode. Finally, observe that a pulse frequency of 1 kHz corresponds to a period (PER) of 1 millisecond (ms). Run a time (transient) analysis of 2 milliseconds duration and perform the following steps:

1. Obtain a Probe plot of both Vpulse and Vout.

2. For V1 = 4 volts, what is the voltage of Vout?

3. Does this voltage differ if the diode used was an ideal diode?

4. For V1 = −4 volts, what is the voltage of Vout?

5. Does this voltage differ if the diode used was an ideal diode?

6. Compare your simulation results with those obtained in the laboratory.

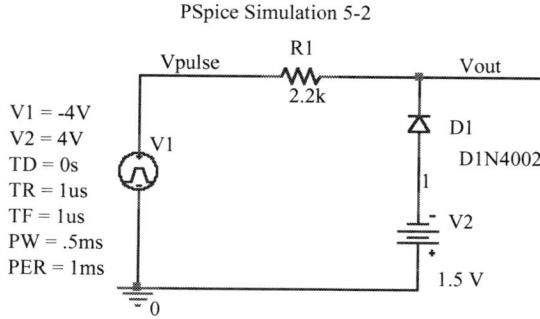

PSpice Simulation 5-2

7. Comment on the agreement or discrepancy between the two sets of data.

8. Reverse the diode and repeat the analysis.

PSpice Simulation 5-3

The circuit in this simulation is that of the series clipper circuit of Fig. 5.11. The parameters of the voltage source V2 are set as in the previous simulation. Run a Time Domain (transient) analysis of 2 milliseconds duration and perform the following steps:

1. Obtain a Probe plot of Vpulse and Vout.

2. Compare this plot to the data obtained experimentally for the interval of both the positive and the negative portions of the input voltage Vpulse.

Exp. 5 / Clipping Circuits

3. Is there a significant difference in the two sets of data?

4. Explain the "clipping" action of this circuit from that data.

5. Obtain a plot of the diode voltage versus the input voltage.

6. What is the diode voltage during the positive portion of Vpulse?

7. Compare this voltage with the transition voltage of that diode.

8. What is the diode voltage during the negative portion of Vpulse?

9. Explain the difference in voltage between the "on" state and the "off" state of the diode.

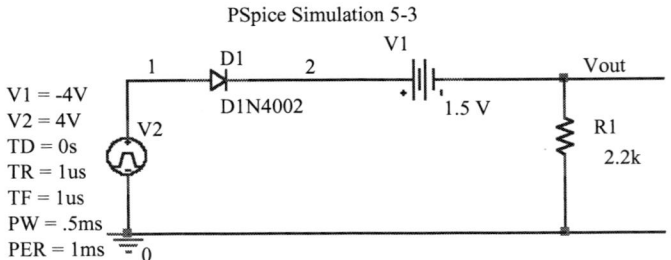

PSpice Simulation 5-3

Name _____
Date _____
Instructor _____

Clamping Circuits

OBJECTIVE

To calculate, draw, and measure the output voltage of clampers.

EQUIPMENT REQUIRED

Instruments

Oscilloscope
DMM

Components

Resistors

(1) 100-Ω
(1) 1-kΩ
(1) 100-kΩ

Diode

(1) Silicon

Capacitor

(1) 1-μF

Supplies

(1) 1.5-V D cell and holder
Function generator

Exp. 6 / Clamping Circuits

EQUIPMENT ISSUED

Item	Laboratory serial no.
Oscilloscope	
DMM	
Function generator	

RÉSUMÉ OF THEORY

Clampers are designed to "clamp" an alternating input signal to a specific level without altering the peak-to-peak characteristics of the waveform. Clampers are easily distinguished from clippers in that they include a capacitive element. A typical clamper will include a capacitor, diode, and resistor, with some also having a DC battery. The best approach to the analysis of clampers is to use a step-by-step approach. The first step should be an examination of the network for that part of the input signal that forward biases the diode. Choosing this part of the input signal will save time and probably avoid some unnecessary confusion. With the diode forward biased the voltage across the capacitor and across the output terminals can be determined. For the rest of the analysis it is then assumed that the capacitor will hold on to the charge and voltage level established during this interval of the input signal. The next part of the input signal can then be analyzed to determine the effect of the stored voltage across the capacitor and the open-circuit state of the diode on the output voltage.

The analysis of a clamper can be quickly checked by simply noting whether the peak-to-peak voltage of the output signal is the same as the peak-to-peak voltage of the applied signal. It is a characteristic of clampers that must be satisfied.

PROCEDURE

Part 1. Threshold Voltage

Determine the threshold voltage for the silicon diode using the diode-checking capability of the DMM or a curve tracer. If either approach is unavailable assume $V_T = 0.7$ V.

$V_T =$ _____

Part 2. Clampers (R, C, Diode Combination)

a. Construct the network of Fig. 6.1 and record the measured value of R.

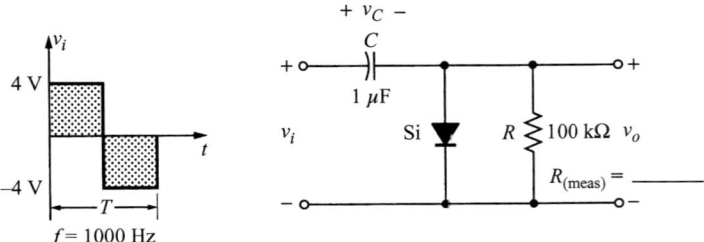

Figure 6-1

Exp. 6 / Procedure

b. Using the value of V_T from Part 1, calculate V_C and V_o for the interval of v_i that causes the diode to be in the "on" state.

V_C (calculated) = _____

V_o (calculated) = _____

c. Using the results of Part 2(**b**), calculate the level of V_o after v_i switches to the other level and turns the diode "off."

V_o (calculated) = _____

d. Using the results of Parts 2(**b**) and 2(**c**), sketch the expected waveform for V_o in Fig. 6.2 for one full cycle of V_i. Use the horizontal center axis as the $V_o = 0$ V line. Record the chosen vertical and horizontal sensitivities below:

Sketch of V_o from calculated results:

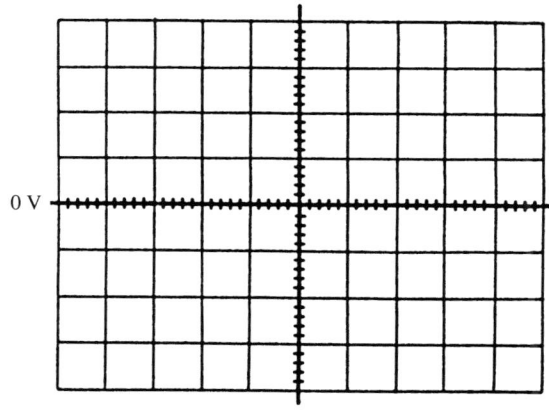

Figure 6-2

Vertical sensitivity = _____

Horizontal sensitivity = _____

e. Using the sensitivities of Part 2(**b**), use the oscilloscope to view the output waveform v_o. Be sure to preset the $V_o = 0$ V line on the screen using the GND position of the coupling switch (and the DC position to view the waveform). Record the resulting waveform on Fig. 6.3.

How does the waveform of Fig. 6.3 compare with the expected waveform of Fig. 6.2?

Sketch of V_o from measured results:

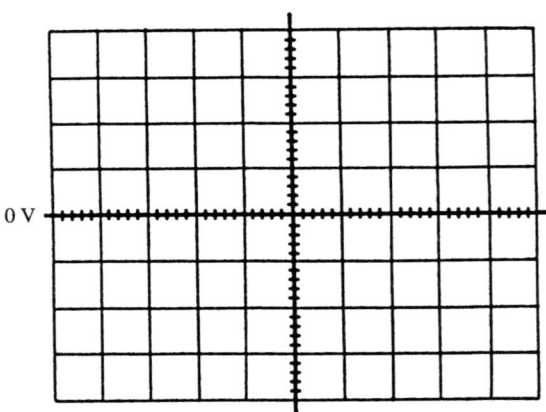

Figure 6-3

f. Reverse the diode of Fig. 6.1 and, using the value of V_T from Part 1, determine the levels of V_C and V_o for the interval of v_i that causes the diode to be in the "on" state.

V_C (calculated) = _____
V_o (calculated) = _____

g. Using the results of Part 2(**f**), calculate the level of V_o after v_i switches to the other level and turns the diode "off."

V_o (calculated) = _____

h. Using the results of Parts 2(**f**) and 2(**g**), sketch the expected waveform for v_o on Fig. 6.4. Use the horizontal axis as the $v_o = 0$ V line. Record the chosen vertical and horizontal sensitivities below:

Sketch of V_o from calculated response:

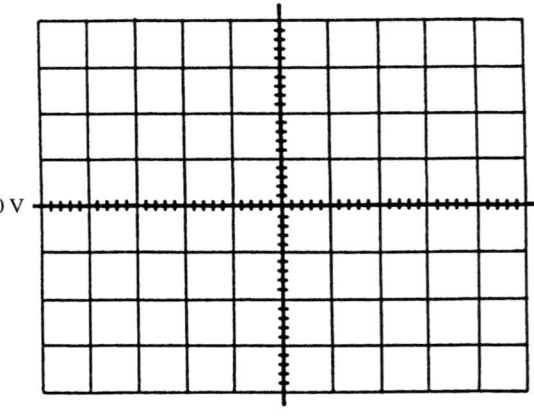

Figure 6-4

Vertical sensitivity = _____
Horizontal sensitivity = _____

i. Using the sensitivities of Part 2(**h**), use the oscilloscope to view the output waveform v_o. Be sure to preset the $V_o = 0$ V line on the screen using the GND position of the coupling switch (and the DC position to view the waveform). Record the resulting waveform on Fig. 6.5.

Sketch of V_o from measured results:

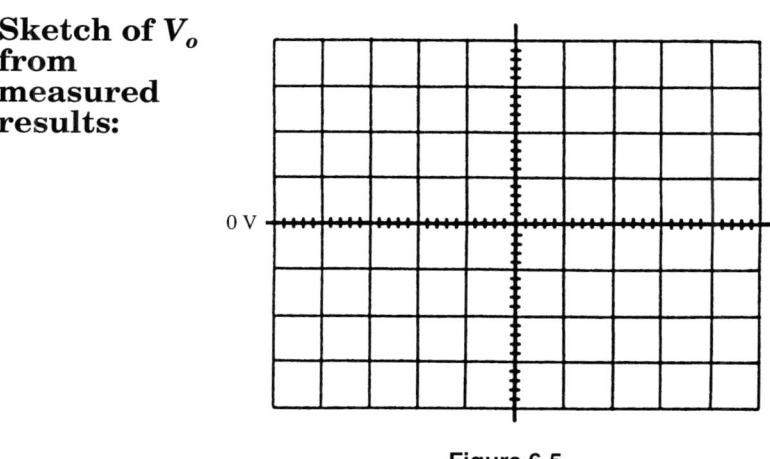

Figure 6-5

How does the waveform of Fig. 6.5 compare with the expected waveform of Fig. 6.4?

Part 3. Clampers with a DC Battery

a. Construct the network of Fig. 6.6 and record the measured values of R and E.

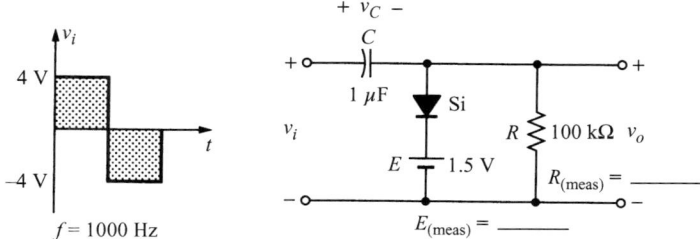

Figure 6-6

b. Using the value of V_T from Part 1, calculate V_C and v_o for the interval of v_i that causes the diode to be in the "on" state.

V_C (calculated) = _____

V_o (calculated) = _____

c. Using the results of Part 3(**b**), calculate the level of v_o after v_i switches to the other level and turns the diode "off."

V_o (calculated) = _____

d. Using the results of Parts 3(**b**) and 3(**c**), sketch the expected waveform for v_o on Fig. 6.7. Use the horizontal center axis as the $V_o = 0$ V line. Record the chosen vertical and horizontal sensitivities below:

Sketch of V_o from calculated results:

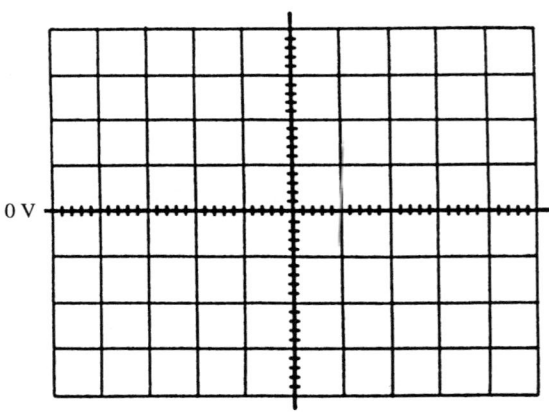

Figure 6-7

Vertical sensitivity = _____

Horizontal sensitivity = _____

e. Using the sensitivities of Part 3(**d**), use the oscilloscope to view the output waveform v_o. Be sure to preset the $V_o = 0$ V line on the screen using the GND position of the coupling switch (and the DC position to view the waveform). Record the resulting waveform on Fig. 6.8.

How does the waveform of Fig. 6.8 compare with the expected waveform of Fig. 6.7?

Sketch of V_o from measured results:

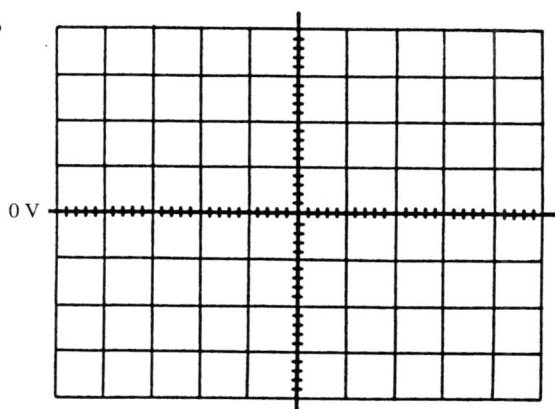

Figure 6-8

f. Reverse the diode of Fig. 6.6 and, using the value of V_T from Part 1, calculate the levels of V_C and V_o for the interval of the input voltage v_i that causes the diode to be in the "on" state.

V_C (calculated) = _____
V_o (calculated) = _____

g. Using the results of Part 3(**f**), calculate the level of V_o after v_i switches to the other level and turns the diode "off."

V_o (calculated) = _____

h. Using the results of Parts 3(**f**) and 3(**g**), sketch the expected waveform for v_o on Fig. 6.9. Use the horizontal center axis as the $V_o = 0$ V line. Record the chosen vertical and horizontal sensitivities below:

Sketch of V_o from calculated results:

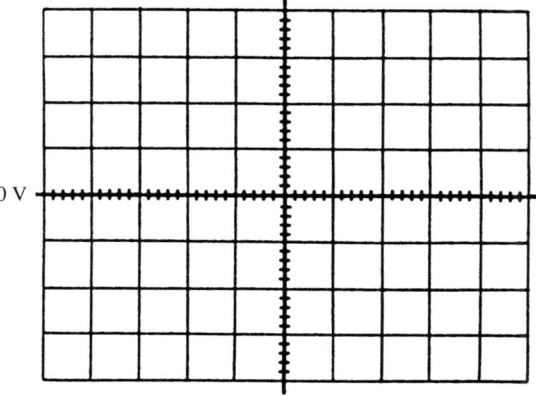

Figure 6-9

Vertical sensitivity = _____
Horizontal sensitivity = _____

i. Using the sensitivities of Part 3(**h**), use the oscilloscope to view the output waveform v_o. Be sure to preset the $V_o = 0$ V line on the screen using the GND position of the coupling switch (and the DC position to view the waveform). Record the resulting waveform on Fig. 6.10.

Sketch of V_o from measured results:

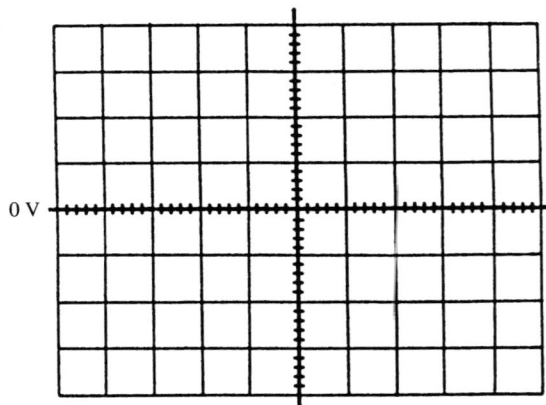

Figure 6-10

How does the waveform of Fig. 6.10 compare with the expected waveform of Fig. 6.9?

Part 4. Clampers (Sinusoidal Input)

a. Reconstruct the network of Fig. 6.1 but change the input signal to an 8 V_{p-p} sinusoidal signal with the same frequency (1000 Hz).

b. Using the results of Parts 1 and 2 and any other analysis technique at your disposal, sketch the expected output waveform for v_o on Fig. 6.11. In particular, find v_o when v_i is its positive and negative peak value and when $v_i = 0$ V. Record the chosen vertical and horizontal sensitivities below:

Sketch of V_o from calculated results:

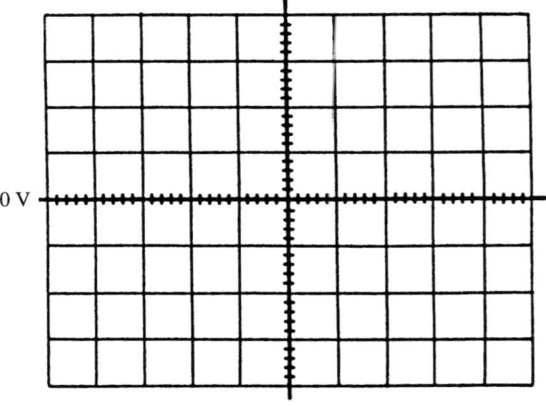

Figure 6-11

Exp. 6 / Procedure   75

V_o (calculated) when $V_i = +4$ V is _____
V_o (calculated) when $V_i = -4$ V is _____
V_o (calculated) when $V_i = 0$ V is _____
Vertical sensitivity = _____
Horizontal sensitivity = _____

c. Using the sensitivities of Part 4(**b**), use the oscilloscope to view the output waveform v_o. Be sure to preset the $V_o = 0$ V line on the screen using the GND position of the coupling switch (and the DC position to view the waveform). Record the resulting waveform on Fig. 6.12.

Sketch of V_o from measured results:

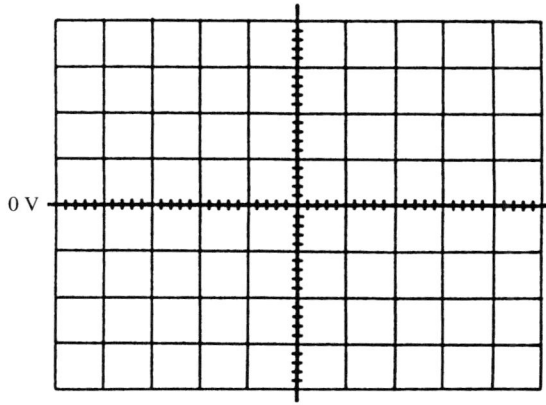

Figure 6-12

How does the waveform of Fig. 6.12 compare with the expected waveform of Fig. 6.11?

Part 5. Clampers (Effect of R)

a. Determine the time constant ($\tau = RC$) for the network of Fig. 6.1 for the interval of the input signal that causes the diode to assume the "off" state and be approximated by an open circuit.

τ (calculated) = _____

b. Calculate the period of the applied signal. Determine half the period to correspond with the time interval that the diode is in the "off" state during the first cycle of the applied signal.

T (calculated) = _____
$T/2$ (calculated) = _____

c. The discharge period of an RC network is about 5τ. Calculate the time interval established by 5τ using the result of Part 5(a) and compare to $T/2$ calculated in Part 5(b).

5τ (calculated) = _____

d. For good clamping action, why is it important for the time interval specified by 5τ to be much larger than $T/2$ of the applied signal?

e. Change R to 1 kΩ and calculate the new value of 5τ.

5τ (calculated) = _____

f. How does the 5τ calculated in Part 5(e) compare to $T/2$ of the applied signal? How would you expect the new value of R to affect the output waveform v_o?

g. Set the input of Fig. 6.1 with R = 1 kΩ and record the resulting waveform on Fig. 6.13. Be sure to preset the V_o = 0 V line in the center of the screen using the GND position of the coupling switch and be sure to use the DC position to view the waveform. Insert the chosen vertical and horizontal sensitivities below:

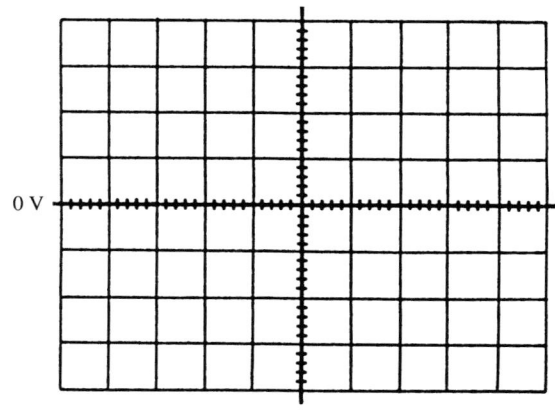

Figure 6-13

Vertical sensitivity = _____
Horizontal sensitivity = _____

h. Comment on the resulting waveform of Fig. 6.13. Is the distortion as you expected? Are you surprised by the positive and negative peaks? Why?

i. Change R to 100 Ω and calculate the new value of 5τ.

5τ (calculated) = _____

j. How does the 5τ calculated in Part 5(i) compare to $T/2$ for the applied signal? What effect will the lower value of R have on the waveform of Fig. 6.13?

k. Set the input of Fig. 6.1 with R = 100 Ω and record the resulting waveform on Fig. 6.14. Be sure to preset the V_o = 0 V line using the coupling switch and use the DC position to view the waveform v_o. Insert the chosen vertical and horizontal sensitivities below:

Vertical sensitivity = _____
Horizontal sensitivity = _____

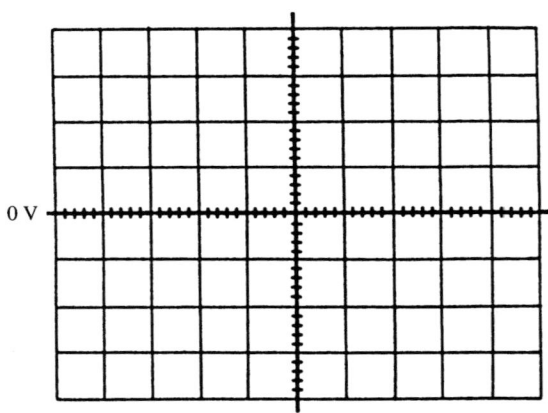

Figure 6-14

l. Comment on the resulting waveform of Fig. 6.14. Compare it to the waveform of Fig. 6.13 and the properly clamped waveform of Fig. 6.3.

m. Using the results of Parts 5(a) through 5(l), establish a relationship between 5τ and the period of the waveform (T) that will ensure that the output waveform has the same characteristics as the input. Note that the requested relationship is between 5τ and T and not $T/2$.

Part 6. Computer Exercises

PSpice Simulation 6-1

1. Construct the network of Fig. 6.1 using PSpice. Connect the Vpulse source into the circuit and set its parameters the same as in PSpice Simulation 6-2. Obtain the voltages V_C and V_o. Compare them to the results of Part 2.

2. Repeat the analysis of Part 5(**a**) but with $R = 1$ kΩ. Comment on the resulting shape of the v_o curve.

3. Repeat the analysis of Part 5(**a**) but with $R = 100$ Ω. Comment on the resulting shape of the v_o curve.

PSpice Simulation 6-2

The circuit used in this simulation is that of Fig. 6.6. The voltage source parameters of source Vpulse are set as shown. Run a Time Domain (transient) analysis of 2 milliseconds duration and perform the following steps:

1. Obtain the Probe plots of both Vpulse, or the nodal voltage V(1), and the nodal voltage V(2), which is the output voltage of this clamper.

Exp. 6 / Clamping Circuits

2. Compare the shift of the DC level of V(2) obtained in this simulation with that of v_o obtained experimentally.

3. Compare the swing of the voltages Vpulse and V(2). Do they agree with the experimental data obtained?

4. Obtain Probe plots of Vpulse, or V(1), and V(1,2).

5. Explain the relationship between those two voltages.

6. What is the appearance of the capacitor voltage right after Vpulse goes positive for $t < .1$ millisecond?

7. From a Probe plot of the diode voltage V(2,3), what is its PIV value?

PSpice Simulation 6-2: Clamper with DC Battery

Nodal voltage
V(1) = Vpulse.

V(1, 2) = capacitor voltage

C1 = 1uF

D1 = D1N4002

V1 = 1.5 V

R1 = 100k

V2 = Vout

V1 = -4V
V2 = 4V
TD = 0s
TR = 1us
TF = 1us
PW = .5ms
PER = 1ms

Name _____
Date _____
Instructor _____

EXPERIMENT 7

Light-Emitting and Zener Diodes

OBJECTIVE

To calculate, draw, and measure the currents and voltages of light-emitting diodes (LEDs) and Zener diodes.

EQUIPMENT REQUIRED

Instruments

DMM

Components

Resistors

(1) 100-Ω
(1) 220-Ω
(1) 330-Ω
(1) 2.2-kΩ
(1) 3.3-kΩ
(2) 1-kΩ

Diode

(1) Silicon
(1) LED
(1) Zener (10-V)

Supplies

DC power supply

EQUIPMENT ISSUED

Item	Laboratory serial no.
DMM	
DC power supply	

RESUMÉ OF THEORY

The light-emitting diode (LED) is, as the name implies, a diode that will give off visible light when sufficiently energized. In any forward-biased *p-n* junction there is, close to the junction, a recombination of holes and electrons. This recombination requires that the energy possessed by unbound free electrons be transferred to another state. In LED materials, such as gallium arsenide phosphide (GaAsP) or gallium phosphide (GaP), photons of light energy are emitted in sufficient numbers to create a visible light source—a process referred to as *electroluminescence*. For every LED there is a distinct forward voltage and current that will result in a bright, clear light, whether it be red, yellow, or green. The diode may, therefore, be forward biased, but until the distinct level of voltage and current is reached, the light may not be visible. In this experiment the characteristics of an LED will be plotted and the "firing" levels of voltage and current determined.

The Zener diode is a *p-n* junction device designed to take full advantage of the Zener breakdown region. Once the reverse-bias potential reaches the Zener region, the ideal Zener diode is assumed to have a fixed terminal voltage and zero internal resistance. All practical diodes have some internal resistance even though, typically, it is limited to 5 to 20 Ω. The internal resistance is the source of the variation in Zener voltage with current level. The experimental procedure will demonstrate the variation in terminal voltage for different loads and resulting current levels.

The following procedure is used to determine the state of the Zener diode. For most configurations, the state of the Zener diode can usually be determined simply by replacing the Zener diode with an open circuit and calculating the voltage across the resulting open circuit. If the open-circuit voltage equals or exceeds the Zener potential, the Zener diode is "on" and the Zener diode can be replaced by a DC supply equal to the Zener potential. Even though the open-circuit voltage may be greater than the Zener potential, the diode is still replaced by a supply equal to the Zener potential. Once the Zener voltage is substituted, the remaining voltages and currents of the network can be determined.

PROCEDURE

Part 1. LED Characteristics

a. Construct the circuit of Fig. 7.1. Initially, set the supply voltage to 0 V and record the measured value of the resistor R.

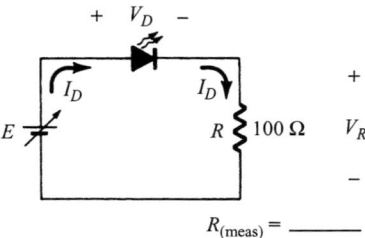

$R_{(meas)} = $ _____

Figure 7-1

b. Increase the supply voltage E until "first light" is noticed. Record the value of V_D and V_R using the DMM. Calculate the corresponding level of I_D using $I_D = V_R/R$ and the measured resistance value.

V_D (measured) = _____
V_R (measured) = _____
I_D (calculated) = _____

c. Continue to increase the supply voltage E until "good brightness" is first established. Don't overload (add too much current to) the circuit and possibly damage the LED by continuing to raise the voltage beyond this level. Record the values of V_D and V_R and calculate the corresponding level of I_D using $I_D = V_R/R$ and the measured resistance value.

V_D (measured) = _____
V_R (measured) = _____
I_D (calculated) = _____

d. Set the DC supply to the levels appearing in Table 7.1 and measure both V_D and V_R. Record the values of V_D and V_R in Table 7.1 and calculate the corresponding level of I_D using $I_D = V_R/R$ and the measured resistance value.

TABLE 7.1

E (V)	0	1	2	3	4	5	6
V_D (V)							
V_R (V)							
$I_D = V_R/R$ (mA)							

e. Using the data of Table 7.1, sketch the curve of I_D vs. V_D on the graph of Fig. 7.2.

f. Draw a light dashed horizontal line across the graph of Fig. 7.2 at the current I_D required for "good brightness." In addition, draw a light dashed vertical line the full height of Fig. 7.2 at the point of intersection between the curve and the light dashed horizontal line. The intersection of the vertical line with the horizontal axis should result in a level of V_D close to that measured in Part 1(c).

Shade in the region below the I_D line and to the left of the V_D line and label the region as the region to be avoided if "good brightness" is to be obtained. Label the remaining unshaded region of Fig. 7.2 as the region for "good brightness."

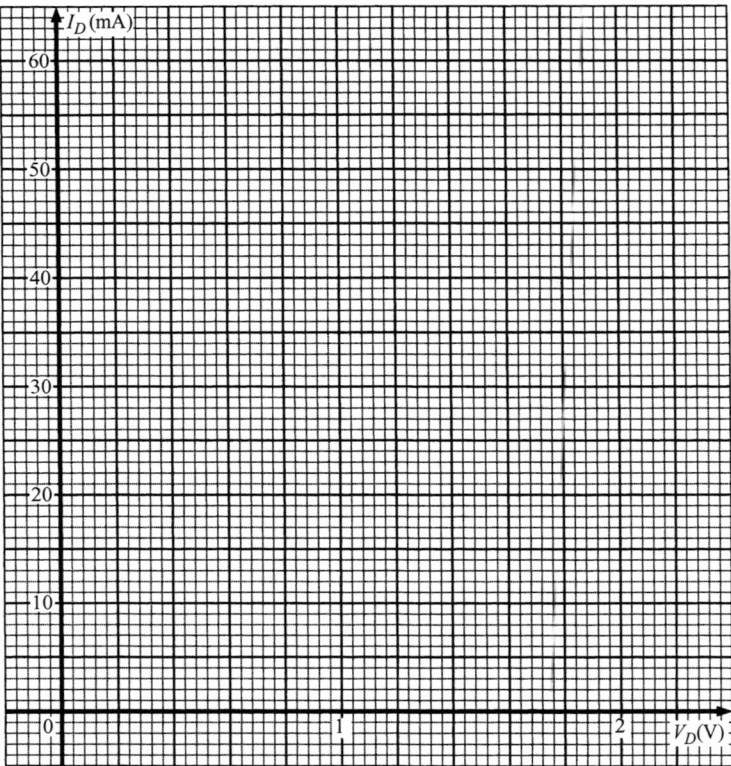

Figure 7-2

g. Construct the circuit of Fig. 7.3. Be sure that both diodes are connected properly and record the measured resistance value.

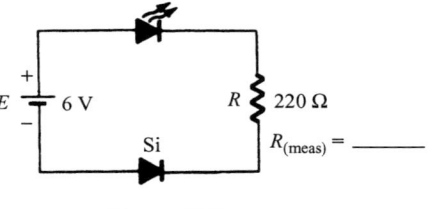

Figure 7-3

h. Do you expect the LED to burn brightly? Why?

i. Energize the network of Fig. 7.3 and verify your conclusion in Part 1(**h**).

j. Reverse the silicon diode of Fig. 7.3 and repeat Part 1(**h**).

k. Repeat Part 1(i). If the LED is "on" with "good brightness," measure V_D and V_R and calculate the level of I_D. Find the intersection of I_D and V_D on the graph of Fig. 7.2. Is the intersection on the curve part of the "good brightness" region?

Part 2. Zener Diode Characteristics

a. Construct the circuit of Fig. 7.4. Initially, set the DC supply to 0 V and record the measured value of R.

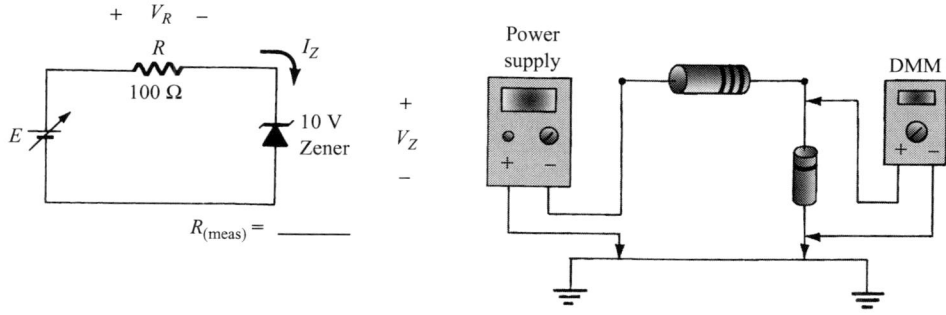

Figure 7-4

b. Set the DC supply (E) to the values appearing in Table 7.2 and measure both V_Z and V_R. You may have to use the millivolt range of your DMM for low values of V_Z and V_R.

TABLE 7.2

E (V)	0	1	2	3	4	5	6	7	8	9	10	11	12	13	14	15
V_Z (V)																
V_R (V)																
$I_Z = V_R/R_{meas}$ (mA)																

c. Calculate the Zener current I_Z in mA at each level of E using Ohm's law as indicated in the last row of Table 7.2 and complete the table.

d. This step will develop the characteristic curve for the Zener diode. Since the Zener region is in the third quadrant of a complete diode characteristic curve, place a minus sign in front of each level of I_Z and V_Z for each data point. With this convention in mind plot the data of Table 7.2 on the graph of Fig. 7.5. Choose an appropriate scale for I_Z and V_Z as determined by the range of values for each parameter.

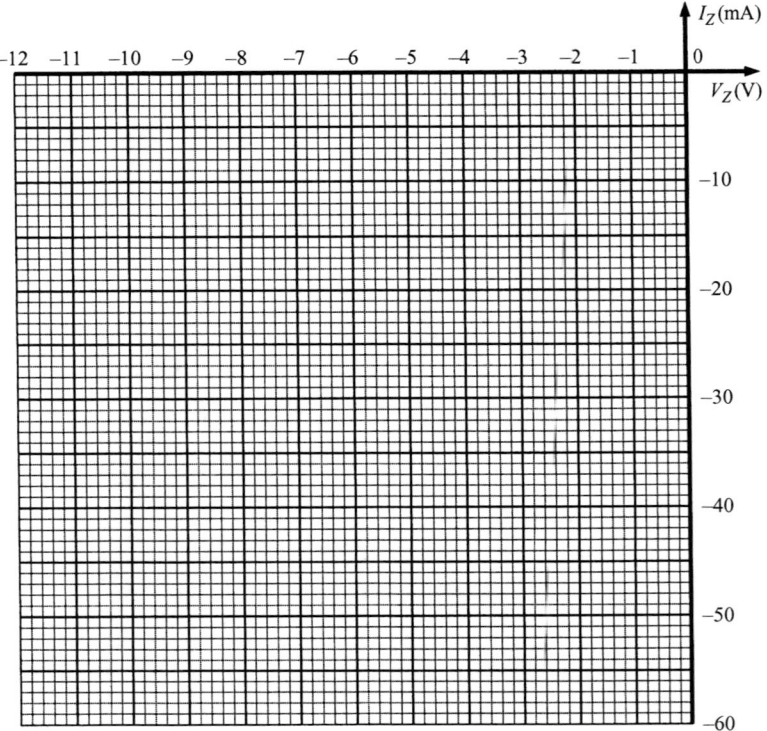

Figure 7-5

e. For the range of measurable current I_Z in the linear (straight line) region that drops from the V_Z axis, what is the average value of V_Z?

V_Z (approximated) = _____

f. For the range of measurable current I_Z in the linear region that drops from the V_Z axis, estimate the average resistance of the Zener diode using $r_{av} = \Delta V_Z/\Delta I_Z$, where ΔV_Z is the change in Zener voltage for the corresponding change in Zener current. Choose an interval of at least 20 V on the linear region of the curve. If necessary, use the data of Table 7.2. Show all work.

R_Z (calculated) = _____

g. Using the results of Parts 2(e) and 2(f), establish the Zener diode equivalent circuit of Fig. 7.6 for the "on" linear region. That is, insert the values of R_Z and V_Z.

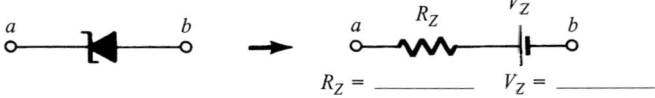

R_Z = _____ V_Z = _____

Figure 7-6

h. For the region from V_Z and $I_Z = 0$ to the point where the characteristic curve drops sharply from the V_Z axis, calculate the resistance of the Zener diode using the equation $r = \Delta V_Z/\Delta I_Z$. Choose $\Delta V_Z = V_Z - 0 \text{ V} = V_Z$ and substitute the resulting change in current (ΔI_Z) for this interval.

R_Z (calculated) = _____

Is the calculated level the level you expected for the region in which the Zener diode is "off"? What would be an appropriate approximation for the Zener diode in this region?

Part 3. Zener Diode Regulation

a. Construct the network of Fig. 7.7. Record the measured value of each resistor.

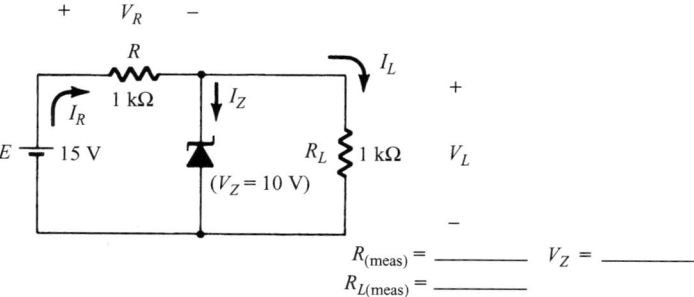

$R_{(meas)}$ = _____ V_Z = _____
$R_{L(meas)}$ = _____

Figure 7-7

b. Determine whether the Zener diode of Fig. 7.7 is in the "on" state, that is, operating in the Zener breakdown region. Use the measured resistor values and the V_Z determined in Part 2(e). Ignore the effects of R_Z in your calculations. For the diode in the "on" state calculate the expected values of V_L, V_R, I_R, I_L, and I_Z. Show all calculations.

V_L (calculated) = _____
V_R (calculated) = _____
I_R (calculated) = _____
I_L (calculated) = _____
I_Z (calculated) = _____

c. Energize the network of Fig. 7.7 and measure V_L and V_R. Using these values, calculate the levels of I_R, I_L, and I_Z.

V_L (measured) = _____
V_R (measured) = _____
I_R (calculated) = _____
I_L (calculated) = _____
I_Z (calculated) = _____

How do the results of Parts 3(b) and 3(c) compare?

d. Change R_L to 3.3 kΩ and repeat Part 3(b). That is, calculate the expected levels of V_L, V_R, I_R, I_L, and I_Z using measured resistor values and the V_Z determined in Part 2(e).

Exp. 7 / Procedure

V_L (calculated) = _____
V_R (calculated) = _____
I_R (calculated) = _____
I_L (calculated) = _____
I_Z (calculated) = _____

e. Energize the network of Fig. 7.7 with $R_L = 3.3$ kΩ and $R_1 = 1$ kΩ. Measure V_L and V_R. Using these values, calculate the levels of I_R, I_L, and I_Z.

R_1 (measured) = _____
R_3 (measured) = _____
V_L (measured) = _____
V_R (measured) = _____
I_R (calculated) = _____
I_L (calculated) = _____
I_Z (calculated) = _____

How do the results of Parts 3(d) and 3(e) compare?

f. Using the measured resistor values and V_Z determined from Part 2(e), determine the minimum value of R_L required to ensure that the Zener diode is in the "on" state.

$R_{L_{min}}$ (calculated) = _____

g. Based on the results of Part 3(f), will a load resistor of 2.2 kΩ place the Zener diode of Fig. 7.7 in the "on" state?

Insert $R_L = 2.2$ kΩ into Fig. 7.7 and measure V_L.

V_L (measured) = _____

Are the conclusions of Parts 3(**f**) and 3(**g**) verified?

Part 4. LED-Zener Diode Combination

a. In this part of the experiment we will determine the minimum supply voltage necessary to turn on ("good brightness") the LED and the Zener diode of Fig. 7.8. The LED will reveal when the Zener diode is "on" and the required supply voltage will be the minimum value that can be applied if the Zener diode is to be used to regulate the voltage V_L.

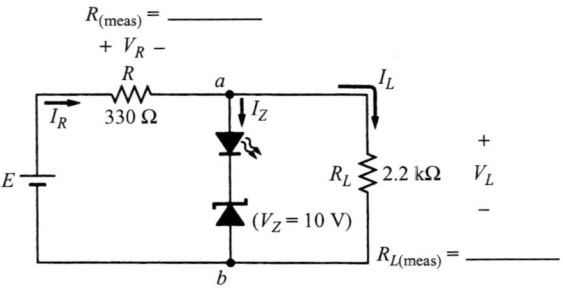

Figure 7-8

b. Refer to Part 1(**c**) and record the level of V_D and I_D that resulted in a "good brightness" level for the LED.

$V_D =$ _____
$I_D =$ _____

Refer to Part 2(**e**) and record the level of V_Z for your Zener diode.

$V_Z =$ _____

Using the above data, determine the total voltage necessary to turn both the LED diode and the Zener diode "on" in Fig. 7.8. That is, determine the required voltage from point a to b.

V_{ab} (calculated) = _____

c. Using the result of Part 4(**b**), calculate the voltage V_L and resulting current I_L. Use measured resistor values.

V_L (calculated) = _____
I_L (calculated) = _____

d. Calculate I_R from $I_R = I_L + I_Z = I_L + I_D$ using the level of I_D from Part 4(**b**). Then calculate the voltage V_R using Ohm's law.

I_R (calculated) = _____
V_R (calculated) = _____

e. Using Kirchhoff's voltage law, calculate the required supply voltage E to turn on the Zener diode and establish "good brightness" by the LED. Use measured resistor values.

E (calculated) = _____

f. Turn on the supply of Fig. 7.8 and increase the voltage E until the LED has "good brightness." Record the required level of E below:

E (measured) = _____

How does the level calculated in Part 4(**e**) compare with the measured value?

g. Measure the voltage V_D and compare with the level listed in Part 4(**b**).

V_D (measured) = _____

Measure the voltage V_Z and compare with the level listed in Part 4(**b**).

V_Z (measured) = _____

Part 5. Computer Exercise

PSpice Simulation 7-1

The regulator shown is similar to that in Fig. 7.7. The diode model D1N750 has a Zener potential of 4.7 volts and a maximum permissible current of 20 mA.

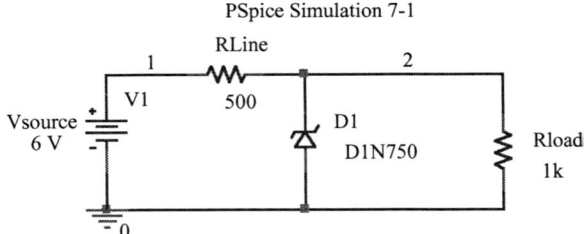

For this circuit perform the following steps:

1. Set the voltage of Vsource to 6 volts and run a bias point analysis.

2. Determine the current through the diode.

3. Determine the voltage across the diode.

4. From this data determine the conducting stage of the diode.

5. Set the voltage of Vsource to 15 volts and repeat the bias point analysis.

Exp. 7 / Part 5. Computer Exercise

6. Determine the current through the diode.

7. Determine the voltage across the diode.

8. From this data determine the conducting stage of the diode.

9. With the operating parameters of this diode, is it safe to operate it given a Vsource of 15 volts?

EXPERIMENT 8

Bipolar Junction Transistor (BJT) Characteristics

OBJECTIVES

1. To determine transistor type (npn, pnp), terminals, and material using a digital multimeter (DMM).
2. To graph the collector characteristics of a transistor using experimental methods and a curve tracer.
3. To determine the value of the alpha and beta ratios of a transistor.

EQUIPMENT REQUIRED

Instruments

DMM
Curve tracer (if available)

Components

Resistors

(1) 1-kΩ
(1) 330-kΩ
(1) 5-kΩ potentiometer
(1) 1-MΩ potentiometer

Transistors

(1) 2N3904 (or equivalent)
(1) Transistor without terminal identification

Supplies

DC power supply

Exp. 8 / Bipolar Junction Transistor (BJT) Characteristics

EQUIPMENT ISSUED

Item	Laboratory serial no.
DMM	
Curve tracer	
DC power supply	

RÉSUMÉ OF THEORY

Bipolar transistors are made of either silicon (Si) or germanium (Ge). Their structure consists of two layers of n-type material separated by a layer of p-type material (npn), or of two layers of p-material separated by a layer of n-material (pnp). In either case, the center layer forms the base of the transistor, while the external layers form the collector and the emitter of the transistor. It is this structure that determines the polarities of any voltages applied and the direction of the electron or conventional current flow. With regard to the latter, the arrow at the emitter terminal of the transistor symbol for either type of transistor points in the direction of conventional current flow and thus provides a useful reference (Fig. 8.2). One part of this experiment will demonstrate how you can determine the type of transistor and its material, and identify its three terminals.

The relationships between the voltages and the currents associated with a bipolar junction transistor under various operating conditions determine its performance. These relationships are collectively known as the characteristics of the transistor. As such, they are published by the manufacturer of a given transistor in a specification sheet. It is one of the objectives of this laboratory experiment to experimentally measure these characteristics and to compare them to their published values.

PROCEDURE

Part 1. Determination of the Transistor's Type, Terminals, and Material

The following procedure will determine the type, terminals, and material of a transistor. The procedure will utilize the diode testing scale found on many modern multimeters. If no such scale is available, the resistance scales of the meter may be used.

    a. Label the transistor terminals of Fig. 8.1 as 1, 2, and 3. Use the transistor without terminal identification for this part of the experiment.

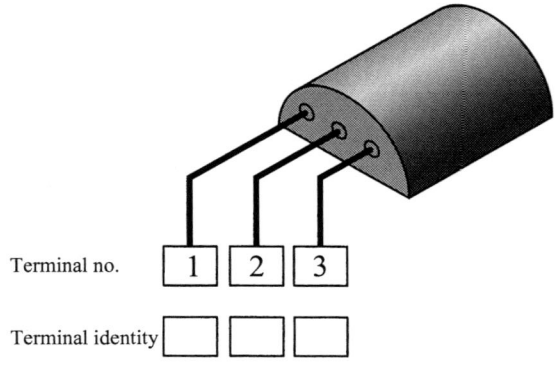

Figure 8-1 Determination of the identities of BJT leads.

b. Set the selector switch of the multimeter to the diode scale (or to the 2 kΩ range if the diode scale is unavailable).

c. Connect the positive lead of the meter to terminal 1 and the negative lead to terminal 2. Record your reading in Table 8.1.

TABLE 8.1

Step	Meter leads connected to BJT		Diode check reading (or highest resistance range)
	Positive	Negative	
c	1	2	
d	2	1	
e	1	3	
f	3	1	
g	2	3	
h	3	2	

d. Reverse the leads and record your reading.

e. Connect the positive lead to terminal 1 and the negative lead to terminal 3. Record your reading.

f. Reverse the leads and record your reading.

g. Connect the positive lead to terminal 2 and the negative lead to terminal 3. Record your reading.

h. Reverse the leads and record your reading.

i. The meter readings between two of the terminals will read high (O.L. or higher resistance) regardless of the polarity of the meter leads connected. Neither of these two terminals will be the base. Based on the above, record the number of the base terminal in Table 8.2.

TABLE 8.2

Part 1 (i):	Base terminal	
Part 1 (j):	Transistor type	
Part 1 (k):	Collector terminal	
Part 1 (k):	Emitter terminal	
Part 1 (l):	Transistor material	

j. Connect the negative lead to the base terminal and the positive lead to either of the other terminals. If the meter reading is low (approximately 0.7 V for Si and 0.3 V for Ge or lower resistance), the transistor type is *pnp;* go to step **k**(1). If the reading is high, the transistor type is *npn;* go to step **k**(2).

k. (1) For *pnp* type, connect the negative lead to the base terminal and the positive lead alternately to either of the other two terminals. The lower of the two readings obtained indicates that the base and collector are connected; thus the other terminal is the emitter. Record the terminals in Table 8.2.

(2) For *npn* type, connect the positive lead to the base terminal and the negative lead alternately to either of the other two terminals. The lower of the two readings obtained indicates that the base and collector are connected; thus the other terminal is the emitter. Record the terminals in Table 8.2.

l. If the readings in either (1) or (2) of Part 1(**k**) were approximately 700 mV, the transistor material is silicon. If the readings were approximately 300 mV, the material is germanium. If the meter does not have a diode testing scale, the material cannot be determined directly. Record the type of material in Table 8.2.

Part 2. The Collector Characteristics

a. Construct the network of Fig. 8.2.

b. Set the voltage V_{R_B} to 3.3 V by varying the 1-MΩ potentiometer. This adjustment will set $I_B = V_{R_B}/R_B$ to 10 µA as indicated in Table 8.3.

c. Then set V_{CE} to 2 V by varying the 5-kΩ potentiometer as required by the first line of Table 8.3.

d. Record the voltages V_{R_C} and V_{BE} in Table 8.3.

e. Vary the 5-kΩ potentiometer to increase V_{CE} from 2 V to the values appearing in Table 8.3. Note that I_B is maintained at 10 µA for the range of V_{CE} levels.

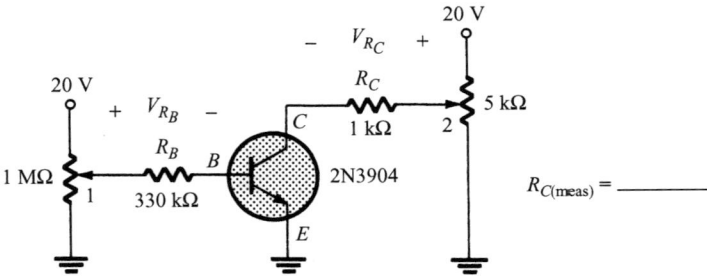

Figure 8-2 Circuit to determine the characteristics of a BJT.

f. For each value of V_{CE} measure and record V_{R_C} and V_{BE}. Use the mV scale for V_{BE}.

g. Repeat Parts 2(**b**) through 2(**f**) for all values of V_{R_B} indicated in Table 8.3. Each value of V_{R_B} will establish a different level of I_B for the sequence of V_{CE} values as shown.

h. After all data have been obtained, compute the values of I_C from $I_C = V_{RC}/R_C$ and I_E from $I_E = I_C + I_B$. Use the measured resistor value for R_C.

i. Using the data of Table 8.3, plot the collector characteristics of the transistor on the graph of Fig. 8.3. That is, plot I_C versus V_{CE} for the various values of I_B. Choose an appropriate scale for I_C and label each I_B curve.

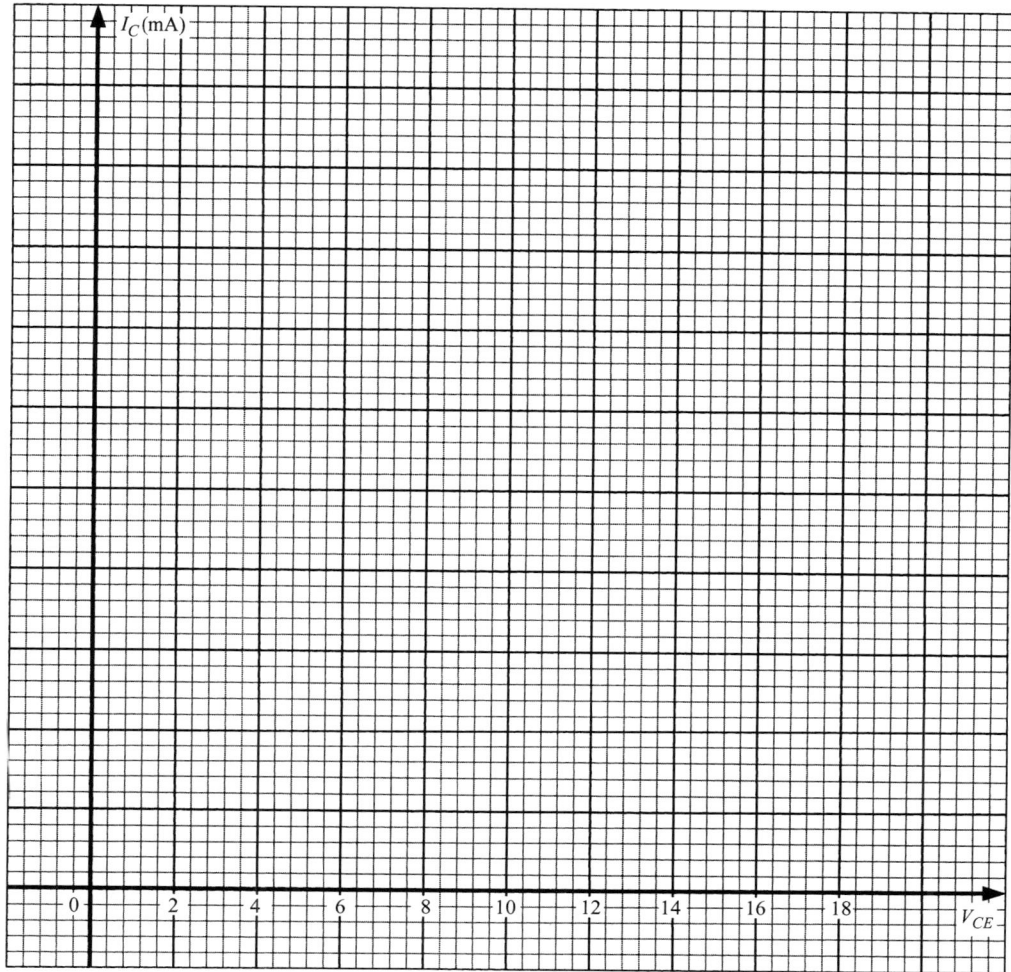

Figure 8-3 Characteristic curves from the experimental data of Part 2.

Part 3. Variation of α and β

a. For each line of Table 8.3 calculate the corresponding levels of α and β using $\alpha = I_C/I_E$ and $\beta = I_C/I_B$ and complete the table.

b. Is there a significant variation in α and β from one region of the characteristics to another?

TABLE 8.3
Data for Construction of Transistor Collector Curve and Calculations of Transistor Parameters

V_{RB} (V) (meas)	I_B (μA) (calc)	V_{CE} (V) (meas)	V_{RC} (V) (meas)	I_C (mA) (calc)	V_{BE} (V) (meas)	I_E (mA) (calc)	α (calc)	β (calc)
↑	↑	2						
│	│	4						
│	│	6						
3.3	10	8						
│	│	10						
│	│	12						
│	│	14						
↓	↓	16						
↑	↑	2						
│	│	4						
│	│	6						
6.6	20	8						
│	│	10						
│	│	12						
↓	↓	14						
↑	↑	2						
│	│	4						
9.9	30	6						
│	│	8						
↓	↓	10						
↑	↑	2						
13.2	40	4						
│	│	6						
↓	↓	8						
↑	↑	2						
16.5	50	4						
↓	↓	6						

In which region are the largest values of β found? Specify using the relative levels of V_{CE} and I_C.

In which region are the smallest values of β found? Specify using the relative levels of V_{CE} and I_C.

c. Find the largest and smallest levels of β and mark their locations on the plot of Fig. 8.3 using the notations $β_{max}$ and $β_{min}$.

d. In general, did β increase or decrease with increase in I_C?

e. In general, did β increase or decrease with increase in V_{CE}? Was the effect of V_{CE} on β greater or less than the effect of I_C?

Part 4. Determination of the Characteristics of a Transistor

Using a Commercial Curve Tracer

a. If available, use a curve tracer to obtain a set of collector characteristics for the 2N3904 transistor. Use the 10 µA step function for I_B and choose a scale for V_{CE} and I_C that matches the scales appearing in the plot of Fig. 8.3.
b. Reproduce the characteristics obtained on the graph of Fig. 8.4. Be sure to label each I_B curve and include the scale for each axis.
c. Compare the characteristics to those obtained in Part 2. Be specific in describing the differences between the two sets of characteristics.

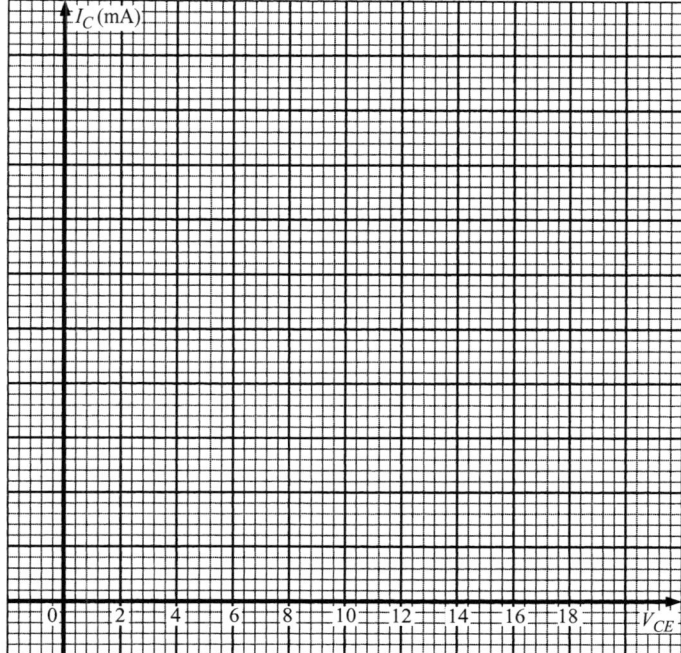

Figure 8-4 Characteristic curves obtained from a commercial curve tracer.

Part 5. Exercises

1. Find the average value of β using the data of Table 8.3. That is, find the sum of the β values and divide by the number of values.

$\beta_{(av)}$ (calculated) = _____

Where on the characteristics did the average value of β typically occur?

Is it reasonable to use this value of β for the transistor for most applications?

2. Determine the average value of V_{BE} using the data of Table 8.3. As in Exercise 1 find the sum of the V_{BE} values and divide by the number of values.

$V_{BE\ (av)}$ (calculated) = _____

Is it reasonable to use the 0.7 V level in the analysis of BJT transistor networks where the actual value is unknown?

3. Careful inspection of the collector curves obtained by experimental measurements and by the curve tracer reveal that the slopes of constant base current are increasing positively (steeper) for higher base currents and higher levels of collector current. What is the effect of the increasing slope of the constant base current lines on the beta of the transistor?

Does the data of Table 8.3 substantiate the above conclusion?

If all the lines of constant base current were horizontal, what would be the effect on the beta ratio determined at any point on a particular base current curve?

If all the lines of constant base current were horizontal and equally spaced, what would be the effect on the beta ratio determined anywhere on the characteristics?

Part 6. Computer Exercise

PSpice Simulation 8-1

1. For the circuit shown, perform a bias point analysis. This circuit corresponds to that of Fig. 8.2 with its two rheostats set to their midpoint resistances. From the simulation data obtained, calculate both the α_{dc} and the β_{dc} of the PSpice model of the Q2N3904 transistor at the operating point of this circuit.

2. Compare the two parameters calculated from the PSpice data with the average value of the same two parameters calculated from the experimental data compiled in Table 8.3.

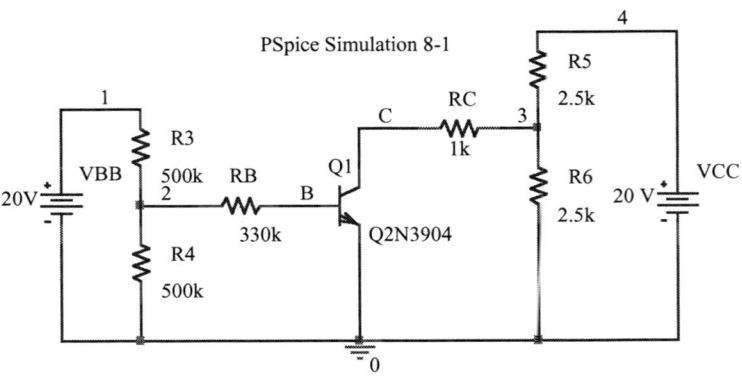

PSpice Simulation 8-1

Fixed- and Voltage-Divider Bias of BJTs

OBJECTIVE

To determine the quiescent operating conditions of the fixed- and voltage-divider-bias BJT configurations.

EQUIPMENT REQUIRED

Instrument

DMM

Components

Resistors

(1) 680-Ω
(1) 2.7-kΩ
(1) 1.8-kΩ
(1) 6.8-kΩ
(1) 33-kΩ
(1) 1-MΩ

Transistors

(1) 2N3904 or equivalent
(1) 2N4401 or equivalent

Supplies

DC power supply

EQUIPMENT ISSUED

Item	Laboratory serial no.
DMM	
DC power supply	

RESUMÉ OF THEORY

Bipolar transistors operate in three modes: cutoff, saturation, and linear. In each of these modes, the physical characteristics of the transistor and the external circuit connected to it uniquely specify the operating point of the transistor. In the cutoff mode, there is only a small amount of reverse current from emitter to collector, making the transistor akin to an open switch. In the saturation mode, there is a maximum current flow from collector to emitter. The amount of that current is limited primarily by the external network connected to the transistor; its operation is analogous to that of a closed switch. Both of these operating modes are used in digital circuits.

For amplification with a minimum of distortion the linear region of the transistor characteristics is employed. A DC voltage is applied to the transistor, forward-biasing the base-emitter junction and reverse-biasing the base-collector junction, typically establishing a quiescent point near or at the center of the linear region.

In this experiment, we will investigate two biasing networks: the fixed-bias and the voltage-divider bias configuration. The former has the serious drawback that the location of the Q-point is very sensitive to the forward current transfer ratio (β) of the transistor and temperature. Because there can be wide variations in beta and the temperature of the device, it can be difficult to predict the exact location of the Q-point on the load line of a fixed-bias configuration.

The voltage-divider bias network employs a feedback arrangement that makes the base-emitter and collector-emitter voltages primarily dependent on the external circuit elements and not the beta of the transistor. Thus, even though the beta of individual transistors may vary considerably, the location of the Q-point on the load line will remain essentially fixed. The phrase "beta-independent biasing" is often used for such an arrangement.

PROCEDURE

Part 1. Determining β

a. Construct the network of Fig. 9.1 using the 2N3904 transistor. Record the measured resistance values.

$R_{B(\text{meas})} = $ _____
$R_{C(\text{meas})} = $ _____

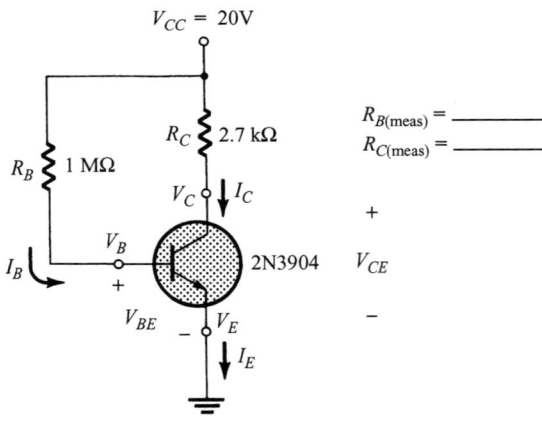

Figure 9-1

b. Measure the voltages V_{BE} and V_{R_C}.

V_{BE} (measured) = _____
V_{R_C} (measured) = _____

c. Using the measured resistor values, calculate the resulting base current using the equation

$$I_B = \frac{V_{R_B}}{R_B} = \frac{V_{CC} - V_{BE}}{R_B} \text{ mA}$$

and the collector current using the equation

$$I_C = \frac{V_{R_C}}{R_C} \text{ mA}$$

The voltage V_{R_B} was not measured directly for determining I_B because of the loading effects of the meter across the high resistance R_B.

Insert the resulting values of I_B and I_C in Table 9.1.

d. Using the results of Part 1(c), calculate the value of β and record in Table 9.1. This value of beta will be used for the 2N3904 transistor throughout this experiment.

$$\beta = \frac{I_C}{I_B}$$

Part 2. Fixed-Bias Configuration

a. Using the β determined in Part 1, calculate the currents I_B and I_C for the network of Fig. 9.1 using the measured resistor values, the supply voltage, and the above measured value for V_{BE}. That is, determine the theoretical values of I_B and I_C using the network parameters and the value of beta.

I_B (calculated) = _____
I_C (calculated) = _____

Exp. 9 / Fixed- and Voltage-Divider Bias of BJTs

How do the calculated levels of I_B and I_C compare to those determined from measured voltage levels in Part 1(c)?

b. Using the results of Part 2(a), calculate the levels of V_B, V_C, V_E, and V_{CE}.

V_B (calculated) = _____
V_C (calculated) = _____
V_E (calculated) = _____
V_{CE} (calculated) = _____

c. Energize the network of Fig. 9.1 and measure V_B, V_C, V_E, and V_{CE}.

V_B (measured) = _____
V_C (measured) = _____
V_E (measured) = _____
V_{CE} (measured) = _____

How do the measured values compare to the calculated levels of Part 2(b)?

Record the measured value of V_{CE} in Table 9.1.

d. The next part of the experiment will essentially be a repeat of a number of the steps above for a transistor with a higher beta. Our goal is to show the effects of different beta levels on the resulting levels of the important quantities of the network. First the beta level for the other transistor, specifically a 2N4401 transistor, must be determined. Remove the 2N3904 transistor from Fig. 9.1 and insert the 2N4401 transistor, leaving all the resistors and voltage V_{CC} as in Part 1. Then measure the voltages V_{BE} and V_{R_C} using the same equations with measured resistor values. Calculate the levels of I_B and I_C. Then determine the level of β for the 2N4401 transistor.

V_{BE} (measured) = _____

V_{R_C} (measured) = _____

I_B (from measured) = _____

I_C (from measured) = _____

β (calculated) = _____

Record the levels of I_B, I_C, and beta in Table 9.1. In addition, measure the voltage V_{CE} and insert in Table 9.1.

TABLE 9.1

Transistor Type	V_{CE} (volts)	I_C (mA)	I_B (µA)	β
2N3904				
2N4401				

e. Using the following equations, calculate the magnitude (ignore the sign) of the percent change in each quantity due to a change in transistors. The fixed-bias configuration has a high sensitivity to changes in beta, as will be reflected by the results. Place the results of your calculations in Table 9.2.

$$\% \Delta\beta = \frac{|\beta_{(4401)} - \beta_{(3904)}|}{|\beta_{(3904)}|} \times 100\% \qquad \% \Delta I_C = \frac{|I_{C(4401)} - I_{C(3904)}|}{|I_{C(3904)}|} \times 100\%$$

$$\% \Delta V_{CE} = \frac{|V_{CE(4401)} - V_{CE(3904)}|}{|V_{CE(3904)}|} \times 100\% \qquad \% \Delta I_B = \frac{|I_{B(4401)} - I_{B(3904)}|}{|I_{B(3904)}|} \times 100\%$$

(9.1)

TABLE 9.2
Percent Changes in β, I_C, V_{CE}, and I_B

%Δβ	%ΔI_C	%ΔV_{CE}	%ΔI_B

Part 3. Voltage-Divider Configuration

a. Construct the network of Fig. 9.2 using the 2N3904 transistor. Insert the measured value of each resistor.

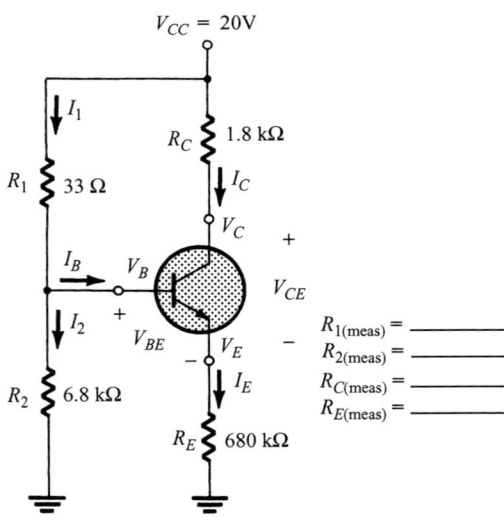

Figure 9-2

b. Using the beta determined in Part 1 for the 2N3904 transistor, calculate the theoretical levels of V_B, V_E, I_E, I_C, V_C, V_{CE}, and I_B for the network of Fig. 9.2. Record the results in Table 9.3.

TABLE 9.3

2N3904	V_B	V_E	V_C	V_{CE}	I_E (mA)	I_C (mA)	I_B (μA)
Calculated [Part 3(**b**)]							
Measured [Part 3(**c**)]							

c. Energize the network of Fig. 9.2 and measure V_B, V_E, V_C, and V_{CE}. Record their values in Table 9.3. In addition, measure the voltages V_{R_1} and V_{R_2}. Try to measure the quantities to the hundredths or thousandths place. Calculate the currents I_E and I_C and the currents I_1 and I_2 (using $I_1 = V_{R_1}/R_1$ and $I_2 = V_{R_2}/R_2$) from the voltage readings and measured resistor values. Using the results for I_1 and I_2, calculate the current I_B using Kirchhoff's current law. Record the calculated current levels for I_E, I_C, and I_B in Table 9.3.

How do the calculated and measured values of Table 9.3 compare? Are there any significant differences that need to be explained?

d. Record the measured value of V_{CE} and calculated values of I_C and I_B from Part 3(**c**) in Table 9.4 along with the magnitude of beta from Part 1.

e. Replace the 2N3904 transistor of Fig. 9.2 with the 2N4401 transistor. Then measure the voltages V_{CE}, V_{R_C}, V_{R_1}, and V_{R_2}. Again, be sure to read V_{R_1} and V_{R_2} to the hundredths or thousandths place to ensure an accurate determination of I_B. Then calculate I_C, I_1, and I_2, and determine I_B. Complete Table 9.4 with the levels of V_{CE}, I_C, I_B, and beta for this transistor.

TABLE 9.4

Transistor Type	V_{CE} (volts)	I_C(mA)	I_B(μA)	β
2N3904				
2N4401				

f. Calculate the percent change in β, I_C, V_{CE}, and I_B from the data of Table 9.4. Use the formulas appearing in Part 2(**e**), Eq. 9.1, and record your results in Table 9.5.

TABLE 9.5
Percent Changes in β, I_C, V_{CE}, and I_B

%Δβ	%ΔI_C	%ΔV_{CE}	%ΔI_B

Part 4. Computer Exercises

PSpice Simulation 9-1

Perform a bias point simulation of the fixed-bias circuit shown. Perform the following steps:

    1. Obtain the base, collector, and emitter currents.

    2. Obtain the base-to-emitter voltage.

    3. Obtain the collector-to-emitter voltage.

Exp. 9 / Part 4. Computer Exercises

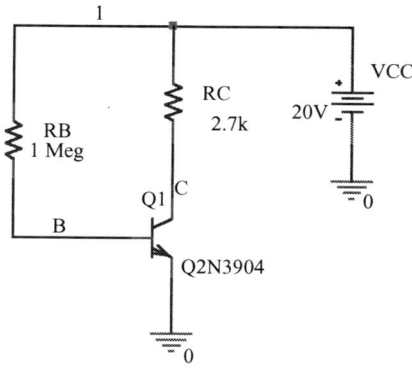

PSpice Simulation 9-1

4. Replace the Q2N3904 transistor with the Q2N2222 transistor of the PSpice program. Note: this transistor does not have as high a beta as does the Q2N4401 transistor used in the experiments. After the replacement, obtain the same data as in the previous simulation.

5. Calculate the percent change in the β of the transistors. In this and all subsequent calculations, use the formulas given in the laboratory text.

6. Calculate the percent changes in the base, collector, and emitter currents.

7. Calculate the percent change in the collector-to-emitter voltage.

8. Calculate the $S(\beta)$ figure of merit from your data.

PSpice Simulation 9-2

Perform a bias point simulation of the voltage-divider circuit shown. Perform the following steps:

1. Obtain the base, collector, and emitter currents.

2. Obtain the base-to-emitter voltage.

3. Obtain the collector-to-emitter voltage.

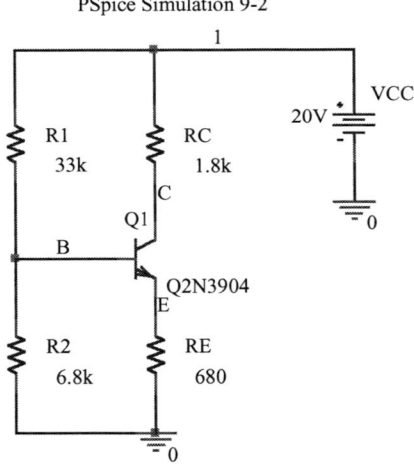

PSpice Simulation 9-2

4. Replace the Q2N3904 transistor with the Q2N2222 transistor of the PSpice program. After the replacement, obtain the same data as in the previous simulation.

5. Calculate the percent change in the β of the transistors.

6. Calculate the percent change in the base, collector, and emitter currents.

7. Calculate the percent change in the collector–to-emitter voltage.

8. Calculate the $S(\beta)$ figure of merit from your data.

Comparison of the fixed-bias with the voltage-divider circuits.

9. Which of the two circuits showed the least variations in the base, collector, and emitter currents when the transistors were changed?

10. Which of the two circuits showed the least variation in the collector-to-emitter voltage when the transistors were changed?

11. Which of the circuits had the smaller $S(\beta)$?

Part 5. Problems and Exercises

1. a. Compute the saturation current $I_{C_{sat}}$ for the fixed-bias configuration of Fig. 9.1.

$I_{C_{sat}}$ (calculated) = _____

b. Compute the saturation current $I_{C_{sat}}$ for the voltage-divider bias configuration of Fig. 9.2.

$I_{C_{sat}}$ (calculated) = _____

c. Are the saturation currents of Exercises 1(**a**) and 1(**b**) sensitive to the beta of the transistor or changes thereof?

2. For both the circuits investigated in this experiment, how did the Q-point location (defined by I_C and V_{CE} on the collector characteristics) change when the 2N3904 transistor was replaced with the 2N4401? That is, how did the Q-point shift location when a transistor with a higher beta was substituted? In particular, did the Q-points move toward saturation (high I_C, low V_{CE}) or cut-off (low I_C, high V_{CE}) conditions?

3. a. Determine the ratio of the change in I_C, V_{CE}, and I_B due to changes in beta and complete Table 9.6. Use the results of Parts 2 and 3 to obtain the percent changes indicated.

TABLE 9.6

	$\dfrac{\%\Delta I_C}{\%\Delta\beta}$	$\dfrac{\%\Delta V_{CE}}{\%\Delta\beta}$	$\dfrac{\%\Delta I_B}{\%\Delta\beta}$
Fixed Bias			
Voltage-Divider			

b. One of the important goals of a good circuit design is to minimize the sensitivity of various circuit currents and voltages to the beta variability of transistors. A figure of merit that quantifies the percent change in collector current for a percent change in beta has been defined by the following equation. In particular, the smaller $S(\beta)$, the less the circuit will be affected by the change in beta.

$$S(\beta) = \frac{\%\Delta I_C}{\%\Delta \beta} \qquad (9.2)$$

Referring to the results of Table 9.6, which network has the better stability factor $S(\beta)$? Is there a significant difference in level between the two stability factors?

c. Do the remaining sensitivities of Table 9.6 support the fact that one configuration is more stable than the other? Which one is more stable?

4. a. For the fixed-bias configuration of Fig. 9.1 develop an equation for I_B in terms of the other elements (voltage source, resistors, β) of the network. Then develop an equation for I_C.

b. Assuming I_1 and I_2 are much larger than I_B, permitting the approximation $I_1 \cong I_2$, develop an equation for I_C in terms of the other elements of the network of Fig. 9.2.

c. Referring to the results of Exercises 4(**a**) and 4(**b**), is there an obvious reason why I_C is more sensitive to changes in beta in one configuration compared to the other?

Name _____
Date _____
Instructor _____

Emitter and Collector Feedback Bias of BJTs

OBJECTIVE

To determine the quiescent operating conditions of the emitter and collector feedback bias BJT configurations.

EQUIPMENT REQUIRED

Instrument

DMM

Components

Resistors

(2) 2.2-kΩ
(1) 3-kΩ
(1) 390-kΩ
(1) 1-MΩ

Transistors

(1) 2N3904 or equivalent
(1) 2N4401 or equivalent

Supplies

DC power supply

Exp. 10 / Emitter and Collector Feedback Bias of BJTs

EQUIPMENT ISSUED

Item	Laboratory serial no.
DMM	
DC power supply	

RÉSUMÉ OF THEORY

This experiment is an extension of Experiment 9. Two additional arrangements will be investigated in this experiment: emitter-bias and collector feedback circuits.

Emitter-Bias Circuit

The emitter-bias configuration in Fig. 10.1 can be constructed using a single or a dual power supply. Both configurations offer increased stability over the fixed bias of Experiment 9. In particular, if the beta of the transistor times the resistance of the emitter resistor is large compared to the resistance of the base resistor, the emitter current becomes essentially independent of the beta of the transistor. Thus, if we exchange transistors in a properly designed emitter-bias circuit, the changes in I_C and V_{CE} should be small.

Collector Feedback Circuit

If we compare the collector feedback bias circuit configuration in Fig. 10.2 with the fixed bias of Experiment 9 it is noted that for the former, the base resistor is connected to the collector terminal of the transistor and not to the fixed supply voltage V_{CC}. Thus the voltage across the base resistance of the collector feedback configuration is a function of the collector voltage and the collector current. In particular, this circuit demonstrates the principle of negative feedback, in which a tendency of an output variable to increase or decrease will result in a reduction or increase in the input variable, respectively. For instance, any tendency on the part of I_C to increase will reduce the level of V_C, which in turn will result in a lower level of I_B offsetting the increasing trend of I_C. The result is a design less sensitive to variations in its parameters.

PROCEDURE

Part 1. Emitter-Bias Configuration: Determining β

a. Construct the network of Fig. 10.1 using the 2N3904 transistor. Insert the measured resistor values.

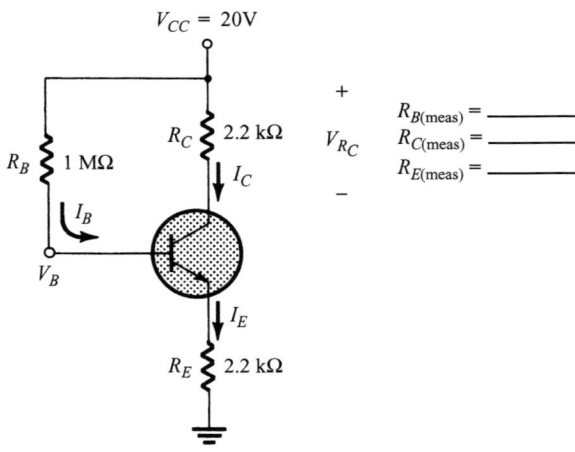

Figure 10-1

Exp. 10 / Procedure

b. Measure the voltages V_B and V_{R_C}.

V_B (measured) = _____

V_{R_C} (measured) = _____

c. Using the results of Part 1(**b**) and the measured resistor values, calculate the resulting base currents I_B and I_C using the following equations:

$$I_B = \frac{V_{CC} - V_B}{R_B} \text{ and } I_C = \frac{V_{R_C}}{R_C}$$

Record in Table 10.2 on page 122.

I_B (calculated from measured) = _____

I_C (calculated from measured) = _____

d. Using the results of Part 1(**c**), calculate the value of β and record in Table 10.2. This value of beta will be used for the 2N3904 transistor throughout the experiment.

β (calculated) = _____

Part 2. Emitter-Bias Configuration: Determining Operating Point

a. Using the β determined in Part 1, calculate the values of I_B and I_C for the network of Fig. 10.1 using measured resistor values and the supply voltage V_{CC}. Perform a theoretical analysis of the network. Insert the results in Table 10.1 on page 122.

I_B (calculated) = _____

I_C (calculated) = _____

How do the calculated values compare with the measured values of Part 1(c)?

b. Using the β determined in Part 1 calculate the levels of V_B, V_C, V_E, V_{BE}, and V_{CE} and insert in Table 10.1.

c. Energize the network of Fig. 10.1 with the 2N3904 and measure the voltages V_B, V_C, V_E, V_{BE}, and V_{CE}. Record in Table 10.2.
How do the calculated and measured results of Tables 10.1 and 10.2 compare for the 2N3904 transistor? In particular, comment on any results that do not compare well.

TABLE 10.1

Transistor Type	Calculated Values						
	V_B (volts)	V_C (volts)	V_E (volts)	V_{BE} (volts)	V_{CE} (volts)	I_B (μA)	I_C (mA)
2N3904							
2N4401							

TABLE 10.2

Transistor Type	Measured Values					(Calc. from Measured Values)		
	V_B (volts)	V_C (volts)	V_E (volts)	V_{BE} (volts)	V_{CE} (volts)	I_B (μA)	I_C (mA)	β
2N3904								
2N4401								

d. Replace the 2N3904 transistor of Fig. 10.1 with the 2N4401 transistor and measure the resulting voltages V_B and V_{R_C}. Then calculate the currents I_B and I_C using measured resistance values. Finally calculate the value of β for this transistor. This will be the value of beta used for the 2N4401 transistor throughout this experiment. Record the levels of I_B, I_C, and β in Table 10.2.

V_B (measured) = _____

V_{R_C} (measured) = _____

e. Using the beta determined in Part 1(**d**), perform a theoretical analysis of Fig. 10.1 with the 2N4401 transistor. Calculate the levels of I_B, I_C, V_B, V_C, V_E, V_{BE}, and V_{CE} and record in Table 10.1.

f. Energize the network of Fig. 10.1 with the 2N4401 transistor; measure V_B, V_C, V_E, V_{BE}, and V_{CE}; and insert in Table 10.2.

How do the calculated and measured results of Tables 10.1 and 10.2 compare for the 2N4401 transistor? Discuss any results that appear different by more than 10%.

g. Calculate the percent change in β, I_C, V_{CE}, and I_B using the equations first presented in Experiment 9 and repeated here for convenience. Record the results in Table 10.3.

$$\% \Delta\beta = \frac{|\beta_{(4401)} - \beta_{(3904)}|}{|\beta_{(3904)}|} \times 100\% \qquad \% \Delta I_C = \frac{|I_{C(4401)} - I_{C(3904)}|}{|I_{C(3904)}|} \times 100\%$$

$$\% \Delta V_{CE} = \frac{|V_{CE(4401)} - V_{CE(3904)}|}{|V_{CE(3904)}|} \times 100\% \qquad \% \Delta I_B = \frac{|I_{B(4401)} - I_{B(3904)}|}{|I_{B(3904)}|} \times 100\%$$

TABLE 10.3
Percent Changes in β, I_C, V_{CE}, and I_B

%Δβ	%ΔI_C	%ΔV_CE	%ΔI_B

Part 3. Collector Feedback Configuration ($R_E = 0\ \Omega$)

a. Construct the network of Fig. 10.2 using the 2N3904 transistor. Record the measured resistor values in Fig. 10.2.

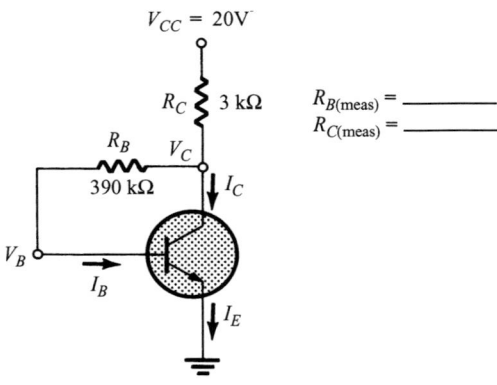

Figure 10-2 Collector Feedback

Exp. 10 / Procedure

b. Using the beta determined in Part 1, calculate the values of I_B, I_C, V_B, V_C, and V_{CE} and record in Table 10.4.

c. Energize the network of Fig. 10.2; measure V_B, V_C, and V_{CE}; and insert in Table 10.5. Calculate the currents I_B and I_C using measured resistance values and the fact that $I_C \cong V_{R_C}/R_C$. Record the current levels in Table 10.5.

How do the calculated and measured results of Tables 10.4 and 10.5 compare for the 2N3904 transistor?

d. Replace the 2N3904 transistor of Fig. 10.2 with the 2N4401 transistor of Part 1; calculate the values of I_B, I_C, V_B, V_C and V_{CE}; and record in Table 10.4.

e. Energize the network of Fig. 10.2 with the 2N4401 transistor and measure V_B, V_C, and V_{CE}. Insert all measurements in Table 10.5. Calculate I_B and I_C from measured values and then record the current levels in Table 10.5.

How do the calculated and measured results of Tables 10.4 and 10.5 compare for the 2N4401 transistor?

f. Calculate the percent changes in β, I_C, V_{CE}, and I_B using the equations of Part 1(**g**). Record the results in Table 10.6.

TABLE 10.4

Transistor Type	Theoretical Calculated Values				
	V_B (volts)	V_C (volts)	V_{CE} (volts)	I_B (μA)	I_C (mA)
2N3904					
2N4401					

TABLE 10.5

Transistor Type	Measured Values			(Calc. from Measured Values)	
	V_B (volts)	V_C (volts)	V_{CE} (volts)	I_B (μA)	I_C (mA)
2N3904					
2N4401					

TABLE 10.6
Percent Changes in β, I_C, V_{CE}, and I_B

%Δβ	%ΔI_C	%ΔV_{CE}	%ΔI_B

Part 4. Collector Feedback Configuration (with R_E)

a. Construct the network of Fig. 10.3 using the 2N3904 transistor. Record the measured resistance values in Fig. 10.3.

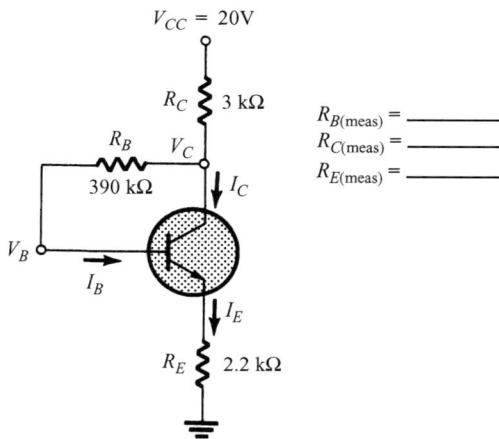

Figure 10-3 Collector feedback circuit.

b. Using the beta determined in Part 1, calculate the values of I_B, I_C, I_E, V_B, V_C, and V_{CE} and record in Table 10.7 on page 129.

c. Energize the network of Fig. 10.3; measure V_B, V_C, V_E, and V_{CE}; and insert in Table 10.8. In addition, calculate the currents I_B, I_C, and I_E from measured values using measured resistor values. Record the current levels in Table 10.8.

How do the calculated and measured results of Tables 10.7 and 10.8 compare for the 2N3904 transistor?

d. Replace the 2N3904 transistor of Fig. 10.3 with the 2N4401 transistor. Using the beta of Part 1, calculate the values of I_B, I_C, I_E, V_B, V_C, and V_{CE} and record in Table 10.7.

e. Energize the network of Fig. 10.3 with the 2N4401 transistor; measure V_B, V_C, V_E, and V_{CE}; and insert in Table 10.8. In addition, calculate the currents I_B, I_C, and I_E from measured values using the measured resistor values. Record the current levels in Table 10.8.

How do the calculated and measured results of Tables 10.7 and 10.8 compare for the 2N4401 transistor?

f. Calculate the percent changes in β, I_C, V_{CE}, and I_B using the equations appearing in Part 1(**g**) and record in Table 10.9.

TABLE 10.7

Transistor Type	Theoretical Calculated Values						
	V_B (volts)	V_C (volts)	V_E (volts)	V_{CE} (volts)	I_B (μA)	I_C (mA)	I_E (mA)
2N3904							
2N4401							

TABLE 10.8

Transistor Type	Measured Values				(Calc. from Measured Values)		
	V_B (volts)	V_C (volts)	V_E (volts)	V_{CE} (volts)	I_B (μA)	I_C (mA)	I_E (mA)
2N3904							
2N4401							

TABLE 10.9
Percent Changes in β, I_C, V_{CE}, and I_B

%Δβ	%ΔI_C	%ΔV_{CE}	%ΔI_B

Part 5. Computer Exercises

PSpice Simulation 10-1

Perform a bias point simulation of the emitter-bias circuit shown. Perform the following steps:

1. Obtain the base, collector, and emitter currents.

2. Obtain the base-to-emitter voltage.

3. Obtain the collector-to-emitter voltage.

4. Obtain the DC power delivered by the source VCC.

5. Obtain the DC power absorbed by each of the resistors.

6. Obtain the DC power absorbed by the transistor.

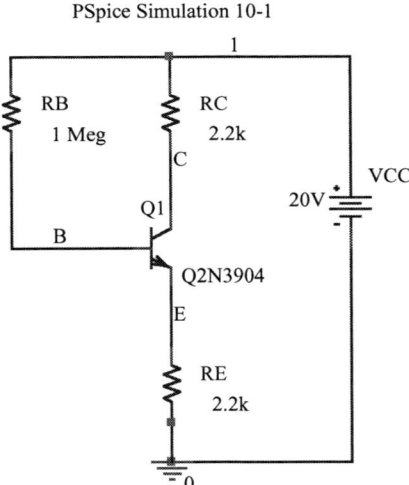

PSpice Simulation 10-1

7. Replace the Q2N3904 transistor with the Q2N2222 transistor of the PSpice program. After the replacement, obtain the same data as in the previous simulation.

8. Calculate the percent change in the β of the transistors.

9. Calculate the percent changes in the base, collector, and emitter currents.

10. Calculate the percent change in the collector-to-emitter voltage.

11. Calculate the $S(\beta)$ figure of merit from your data.

12. Did the DC power delivered by the VCC source change when the transistors were exchanged?

13. Were there any changes in the DC power delivered to the resistors?

14. Was there a change to the DC power supplied to the two transistors?

PSpice Simulation 10-2

Perform a bias point simulation of the collector feedback circuit shown. Perform the following steps:

1. Obtain the base, collector, and emitter currents.

2. Obtain the base-to-emitter voltage.

3. Obtain the collector-to-emitter voltage.

4. Obtain the DC power delivered by the source VCC.

5. Obtain the DC power absorbed by each of the resistors.

6. Obtain the DC power absorbed by the transistor.

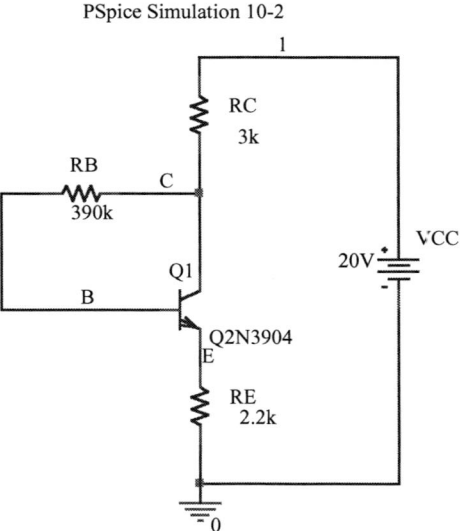

7. Replace the Q2N3904 transistor with the Q2N2222 transistor of the PSpice program. After the replacement, obtain the same data as in the previous simulation.

8. Calculate the percent change in the β of the transistors.

9. Calculate the percent changes in the base, collector, and emitter currents.

10. Calculate the percent change in the collector-to-emitter voltage.

11. Calculate the $S(\beta)$ figure of merit from your data.

12. Did the DC power delivered by the VCC source change when the transistors were exchanged?

13. Were there any changes in the DC power delivered to the resistors?

14. Was there a change to the DC power supplied to the two transistors?

Part 6. Problems and Exercises

1. **a.** Compute the saturation current $I_{C_{sat}}$ for the emitter-bias configuration of Fig. 10.1.

    $I_{C_{sat}}$ (calculated) = _____

    b. Compute the saturation current $I_{C_{sat}}$ for the collector-feedback configuration of Fig. 10.2.

    $I_{C_{sat}}$ (calculated) = _____

    c. Compute the saturation current $I_{C_{sat}}$ for the collector-feedback configuration of Fig. 10.3.

    $I_{C_{sat}}$ (calculated) = _____

    d. What is the effect of beta on the calculations of Exercises 1(**a**), 1(**b**), and 1(**c**)?

2. For the three configurations investigated in this experiment, how did the Q-point location (defined by I_C and V_{CE}) change when the 2N3904 transistor was replaced with the 2N4401? That is, how did the Q-point shift position when a transistor with a higher beta was substituted? In particular, did the Q-points move toward saturation (high I_C, low V_{CE}) or cut-off (low I_C, high V_{CE}) conditions?

3. a. Determine the ratio of the change in I_C, V_{CE}, and I_B due to changes in beta and complete Table 10.10. Use the results of Parts 2, 3, and 4 to obtain the percent changes indicated.

TABLE 10.10

	$\dfrac{\%\Delta I_C}{\%\Delta\beta}$	$\dfrac{\%\Delta V_{CE}}{\%\Delta\beta}$	$\dfrac{\%\Delta I_B}{\%\Delta\beta}$
Emitter bias			
Collector feedback ($R_E = 0\ \Omega$)			
Collector feedback (with R_E)			

b. How does the figure of merit defined by Eq. 9.2 (repeated here for convenience) compare for each configuration of Table 10.10?

$$S(\beta) = \frac{\%\Delta I_C}{\%\Delta\beta}$$

Which appears to have the better stability factor?

c. Do the remaining sensitivities [$S(\beta)$] of Table 10.10 support the fact that one configuration is more stable than the other?

4. a. For the emitter-bias configuration of Fig. 10.1 develop an equation for I_C in terms of the other elements (resistors, V_{CC}, β) of the network. Use the fact that $(\beta + 1) \cong \beta$.

b. Divide the numerator and denominator of the equation obtained in Exercise 4(**a**) by β.

c. Based on the results of Exercise 4(**b**), what relationship must exist between the elements of the network to minimize the effect of changing levels of β on the level of I_C?

5. a. For the collector feedback configuration of Fig. 10.2 develop an equation for I_C in terms of the other elements (resistors, V_{CC}, β) of the network. Use the fact that $(\beta + 1) \cong \beta$.

b. Divide the numerator and denominator of the equation obtained in Exercise 5(**a**) by β.

c. Based on the results of Exercise 5(**b**), what relationship must exist between the elements of the network to minimize the effect of changing levels of β on the level of I_C?

6. a. For the collector feedback configuration of Fig. 10.3 develop an equation for I_C in terms of the other elements (resistors, V_{CC}, β) of the network. Use the fact that $(\beta + 1) \cong \beta$.

b. Divide the numerator and denominator of the equation obtained in Exercise 6(**a**) by β.

c. Based on the results of Exercise 6(**b**), what relationship must exist between the elements of the network to minimize the effect of changing levels of β on the level of I_C?

7. Comparing the results of Exercises 4(c), 5(c), and 6(c), which configuration would appear to have the least sensitivity to changes in beta for resistor values of about the same magnitude?

8. Does the above conclusion compare favorably with the conclusion of Exercise 3(b)?

Name _____
Date _____
Instructor _____

EXPERIMENT 11

Design of BJT Bias Circuits

OBJECTIVE

To design a collector-feedback, emitter-bias, and voltage-divider-bias BJT transistor network.

EQUIPMENT REQUIRED

Instruments

(1) DMM

Components

Resistors

Since this is a design experiment, a number of the required resistors are not specified in the equipment list. They will have to be requested from the stockroom once their values are determined.

(1) 300-Ω, 1.2-kΩ, 1.5-kΩ, 3-kΩ, 15-kΩ, 100-kΩ
(1) 1-MΩ potentiometer
Other resistors as required by the designs

Transistors

(1) 2N3904 or equivalent
(1) 2N4401 or equivalent

Supplies

(1) DC power supply

139

EQUIPMENT ISSUED

Item	Laboratory serial no.
DMM	
DC power supply	

RÉSUMÉ OF THEORY

In this experiment we will make a preliminary design of a collector-feedback, emitter-bias, and voltage-divider-bias BJT transistor configuration. Unlike analysis, where the circuit is given and the response of the circuit variables is asked for, in circuit design, the desired circuit responses are specified and a circuit that yields the desired variables is to be constructed.

Circuit design is often a series of compromises. The most stable network may not result in an acceptable level of AC gain. The resistor values that the theoretical calculations suggest may not be commercially available. One set of resistor values may result in the most stable system with excellent gain characteristics but result in a low conversion efficiency as defined by $\eta\% = P_o(ac)/P_i(dc) \times 100\%$. The designer must be aware of the consequences of making a certain choice and which characteristics of the design are the most vital for the particular application.

For a specified Q-point, β, and an appropriate level for V_E, the following equations can be applied as a DC design sequence:

Collector-Feedback:

$$R_C = \frac{V_{CC} - V_{CE_Q}}{I_{C_Q}} \tag{11.1}$$

$$R_B = \frac{V_{R_B}}{I_B} = \frac{V_{CE_Q} - V_{BE}}{\frac{I_{C_Q}}{\beta}} = \beta\left[\frac{V_{CE_Q} - V_{BE}}{I_{C_Q}}\right] \tag{11.2}$$

Emitter-Bias:

$$R_E = \frac{V_E}{I_{C_Q}} \tag{11.3}$$

$$V_C = V_{CE_Q} + V_E \tag{11.4}$$

$$R_C = \frac{V_{R_C}}{I_{C_Q}} = \frac{V_{CC} - V_C}{I_{C_Q}} \tag{11.5}$$

$$R_B = \frac{V_{R_B}}{I_B} = \frac{V_{CC} - V_{BE} - V_E}{\frac{I_{C_Q}}{\beta}} = \beta\left[\frac{V_{CC} - V_{BE} - V_E}{I_{C_Q}}\right] \tag{11.6}$$

Voltage-Divider Bias:

$$R_C = \frac{V_{CC} - V_{CE_Q} - V_E}{I_{C_Q}} \quad (11.7)$$

$$R_E = \frac{V_E}{I_{C_Q}} \quad (11.8)$$

Assuming $\beta R_E > 10R_2$ will result in

$$V_B = \frac{R_2 V_{CC}}{R_1 + R_2} = V_{BE} + V_E \quad (11.9)$$

Design Criteria

For each of the above configurations the following defines the relative stability of the system:

Collector-Feedback:

$$\text{Increasing } \frac{R_B}{\beta R_C} \text{ decreases stability} \quad (11.10)$$

Emitter-Bias:

$$\text{Increasing } \frac{R_B}{\beta R_E} \text{ decreases stability} \quad (11.11)$$

Voltage-Divider Bias:

$$\text{Increasing } \frac{R_1 \| R_2}{\beta R_E} \text{ decreases stability} \quad (11.12)$$

PROCEDURE

Part 1. Collector-Feedback Configuration

Circuit Specifications:

$$V_{CC} = 15 \text{ V}$$
$$I_{C_Q} = 5 \text{ mA}$$
$$V_{CE_Q} = 7.5 \text{ V}$$

Design Procedure:

a. From the given specifications, determine the required value of R_C for the collector-feedback network of Fig. 11.1.

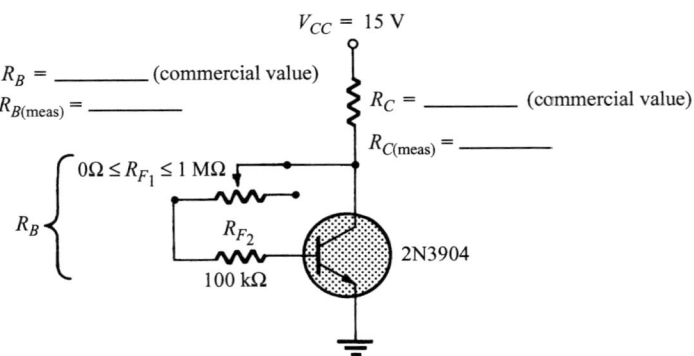

Figure 11-1

Determine the closest commercial value (available in the laboratory) and record below and in Fig. 11.1. Obtain the chosen resistor and insert the measured resistance value in the space provided in Fig. 11.1.

R_C (calculated) = _____
R_C (commercial value) = _____

b. Connect a 100-kΩ resistor and the 1-MΩ potentiometer, set to a maximum, in series, as in Fig. 11.1. Use the commercial value of Part 1 for R_C. With the power on, adjust the potentiometer until V_{CE} = 7.5 V using the 2N3904 transistor.

c. Turn off the supply and disconnect the 100-kΩ resistor from the transistor base connection and measure the combined resistance of R_{F_1} and R_{F_2}. Select a commercial resistor close to this combined level (that is available in the laboratory) and record its nominal value as R_B. In addition, include its measured value on Fig. 11.1.

R_B (measured) = $R_{F_1} + R_{F_2}$ = _____
R_B (commercial value) = _____

d. Replace R_{F_1} and R_{F_2} in the assembled network with the commercial R_B value selected in Part 1(**c**). Then make the measurements and calculations listed below. Use measured values for the resistance levels. Determine I_{C_Q} from $I_{C_Q} = V_{R_C}/R_C$ and I_B from $I_B = (V_{CE} - V_{BE})/R_B$.

V_{R_C} (measured) = _____

V_{CE_Q} (measured) = _____

I_{C_Q} (calculated from measured) = _____

β (calculated) = _____

e. Referring to the results of Part 1(**d**), how do the resulting values of I_{C_Q} and V_{CE_Q} compare to their specified values?

Calculate the percent deviations between all specified and measured values. Use specified values as the standard of comparison.

f. The Résumé of Theory introduced the ratio $\dfrac{R_B}{\beta R_C}$ as an indication of the relative stability of the system. Determine the ratio and insert below. It will be examined in a later part of the experiment.

$R_B/\beta R_C$ (calculated) = _____

g. Rebuild Fig. 11.1 with the 2N4401 transistor and the same value of R_C and repeat Parts 1(**b**) and 1(**c**).

R_{F_1} (measured) + R_{F_2} = _____

R_B (commercial value) = _____

h. Referring to Part 1(**g**), did the value of R_B change with the use of a transistor with a higher β?

Exp. 11 / Design of BJT Bias Circuits

Why did the value of R_C remain the same for both transistors?

i. Repeat Part 1(d) using the 2N4401 transistor.

V_{R_C} (measured) = _____

V_{CE_Q} (measured) = _____

I_{C_Q} (calculated from measured) = _____

β (calculated) = _____

j. Referring to the results of Part 1(i), how do the resulting values of I_{C_Q} and V_{CE_Q} compare to their specified values?

Calculate the percent deviations as in Part 1(e). Which are larger, those presently computed or those in Part 1(e)?

k. Determine the ratio $\dfrac{R_B}{\beta R_C}$ for this configuration and compare to the level calculated for the 2N3904 transistor.

$\dfrac{R_B}{\beta R_C}$ (calculated) (2N4401) = _____

$$\frac{R_B}{\beta R_C} \text{ (calculated) (2N3904)} = \underline{}$$

Compare the results just obtained. What do the results suggest about the two networks when we discuss their relative stability?

l. Rebuild the network of Fig. 11.1 with the level of R_C and R_B calculated for the 2N3904 transistor. However, this time insert the 2N4401 transistor so we can measure the change in I_C due to a change in beta. Energize the network and measure the voltage V_{R_C}. Using the measured resistance value for R_C, calculate I_{C_Q} and then determine $S(\beta) = \frac{\%\Delta I_C}{\%\Delta \beta}$ for the first design.

$S(\beta)$ (calculated) = \underline{}

Part 2. Emitter-Bias Configuration

Circuit Specifications:

$$V_{CC} = 15 \text{ V}$$
$$I_{C_Q} = 5 \text{ mA}$$
$$V_{CE_Q} = 7.5 \text{ V}$$

Design Procedure:

In this case we will employ the design rule that $V_E = 0.1\, V_{CC}$. The Q-point location on the load line is the same as defined for the collector-feedback configuration.

a. For the given specifications, calculate the required value of R_C for the emitter-biased configuration of Fig. 11.2. Determine the closest commercial value (available in the laboratory) and record below and in Fig. 11.2. Obtain the resistor and insert its measured value in the space provided in Fig. 11.2.

R_C (calculated) = \underline{}
R_C (commercial value) = \underline{}

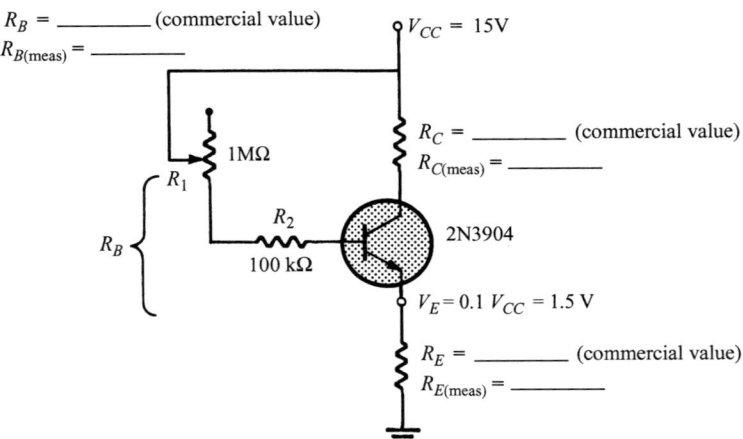

Figure 11-2

b. Using $V_E = 0.1\ V_{CC} = 1.5$ V, calculate the required value of R_E for the network of Fig. 11.2. Determine the closest commercial value (available in the laboratory) and record below and in Fig. 11.2. Obtain the chosen resistor and insert the measured value in the space provided in Fig. 11.2.

R_E (calculated) = _____
R_E (commercial value) = _____

c. Connect a 100-kΩ resistor and the 1-MΩ potentiometer (set to a maximum) in series, as in Fig. 11.2. With the power on, adjust the potentiometer until $V_{CE} = 7.5$ V using the 2N3904 transistor. Use the commercial values of R_C and R_E as determined in Parts 2(**a**) and 2(**b**).

d. Turn off the supply and disconnect R_2 from the base of the transistor. Measure the series resistance defined by $R_1 + R_2$ and record below. Then determine the closest available commercial value (available in the laboratory) and insert below and on Fig. 11.2. Obtain the chosen resistor and record the measured value in Fig. 11.2.

R_B (measured) = $R_1 + R_2$ = _____
R_B (commercial value) = _____

e. Rebuild the network of Fig. 11.2 with the commercial values determined in the above steps. Measure the voltages V_{R_C} and V_{CE} and calculate the current I_C using the measured resistance value

Exp. 11 / Procedure

for R_C. In addition, measure the voltage V_B and calculate the current I_B. Finally, calculate the beta of the transistor.

V_{R_C} (measured) = _____
V_{CE} (measured) = _____
I_C (calculated from measured) = _____
I_B (from measured) = _____
β (calculated) = _____

f. Referring to the results of Part 2(**e**), how do the resulting values of I_{C_Q} and V_{CE_Q} compare to the specified values?

Calculate the percent deviations between all specified and measured values. Use specified values as the standard of comparison.

g. The Résumé of Theory introduced the ratio $R_B/\beta R_E$ as an indication of the relative stability of the system. Determine the ratio and record below.

$R_B/\beta R_E$ (calculated) = _____

h. Rebuild Fig. 11.2 with the 2N4401 transistor and the same values of R_C and R_E and repeat Parts 2(**c**) and 2(**d**).

R_B (calculated) = _____
R_B (commercial value) = _____

i. Referring to Part 1(**h**), did the value of R_B change with the use of the transistor with a higher beta?

Why did the value of R_C and R_E remain the same for both transistors?

j. Repeat Part 2(**e**) using the 2N4401 transistor.

V_{R_C} (measured) = _____
V_{CE_Q} (measured) = _____
I_{C_Q} (calculated from measured) = _____
β (calculated) = _____

k. Referring to the results of Part 2(**j**), how do the resulting values of I_{C_Q} and V_{CE_Q} compare to the specified values? Calculate the percent deviation between the present values and those specified.

l. Since the Q-point for this design is the same for the collector-feedback circuit of Part 1, the magnitudes of beta for each transistor should also be very close for each configuration. Is this conclusion verified by the results of Parts 2(**e**) and 2(**j**)?

m. Determine the ratio $R_B/\beta R_E$ for this configuration and compare to the level calculated for the 2N3904 transistor.

$R_B/\beta R_E$ (2N4401) (calculated) = _____

$R_B/\beta R_E$ (2N3904) (calculated) = _____

Compare the results just obtained. What do the results suggest about the two networks when we discuss their relative stability?

n. Rebuild the network of Fig. 11.2 with the level of R_C, R_E, and R_B calculated for the 2N3904 transistor. However, this time insert the 2N4401 transistor so we can measure the change in I_C due to a change in beta. Energize the network and measure the voltage V_{R_C}. Using the measured resistance value for R_C, calculate I_{C_Q}, and then determine $S(\beta) = \dfrac{\%\Delta I_C}{\%\Delta \beta}$ for the first design.

$S(\beta)$ (calculated) = _____

Part 3. Voltage-Divider Configuration

Circuit Specifications:

$$V_{CC} = 15 \text{ V}$$
$$I_C = 5 \text{ mA}$$
$$V_{CE} = 7.5 \text{ V}$$

Design Procedure:

The Q-point is the same as defined in Parts 1 and 2. In addition, we will continue to apply the rule that $V_E = 0.1\, V_{CC}$.

a. For the given specifications, calculate the required value of R_C for the voltage-divider configuration of Fig. 11.3. Determine the closest commercial value (available in the laboratory) and record below and in Fig. 11.3. Record the measured resistor value in the space provided in Fig. 11.3.

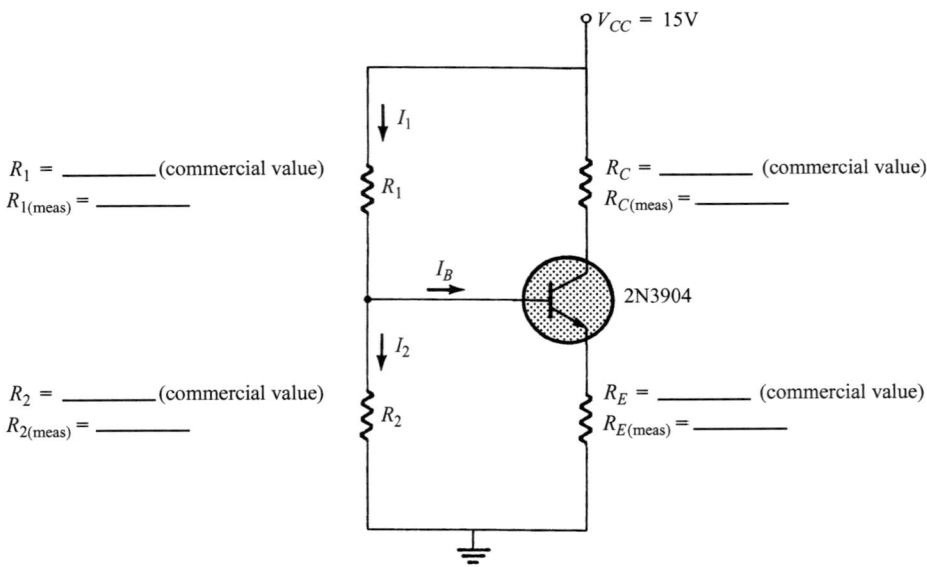

$R_1 = $ _____ (commercial value)
$R_{1(meas)} = $ _____

$R_C = $ _____ (commercial value)
$R_{C(meas)} = $ _____

$R_2 = $ _____ (commercial value)
$R_{2(meas)} = $ _____

$R_E = $ _____ (commercial value)
$R_{E(meas)} = $ _____

Figure 11-3

R_C (calculated) = _____
R_C (commercial value) = _____

b. Using $V_E = 0.1\, V_{CC} = 1.5$ V, calculate the required value of R_E for the network of Fig. 11.3. Determine the closest commercial value (available in the laboratory) and record below and in Fig. 11.3. Insert the measured resistor value in the space provided in Fig. 11.3.

R_E (calculated) = _____

R_E (commercial value) = _____

c. Assuming $\beta R_E > 10R_2$ permits the use of Eq. 11.9 of the Résumé of Theory to define the relationship between R_1 and R_2, determine that relationship using the circuit specifications.

Relationship: _____

d. Using $\beta = 100$ and R_E as determined in Part 3(**b**) (commercial value), calculate the maximum value of R_2 to satisfy the condition $\beta R_E > 10R_2$.

R_2 (calculated) = _____

Choose the nearest standard commercial value (available in the laboratory) for R_2 and calculate the required value of R_1 using the commercial value for R_2 in Eq. 11.9.

R_2 (commercial value) = _____

R_1 (calculated) = _____

Choose the next lowest standard commercial value (available in the laboratory) for R_1 and list below:

R_1 (commercial value) = _____

e. Using the chosen standard commercial values, construct the network of Fig. 11.3 and insert the measured value for each resistor. Measure the voltages V_{R_C} and V_{CE_Q} and calculate the current I_{C_Q} using the measured resistance value for R_C. Finally, make a careful measurement of V_{R_1} and V_{R_2} (at least to the hundredths place) and calculate the currents I_1 and I_2 to the same level of accuracy using measured resistor values. Then determine I_B and calculate β.

$$V_{R_C} \text{ (measured)} = \underline{\hspace{2cm}}$$

$$V_{CE_Q} \text{ (measured)} = \underline{\hspace{2cm}}$$

$$I_{C_Q} \text{ (calculated from measured)} = \underline{\hspace{2cm}}$$

$$\beta \text{ (calculated)} = \underline{\hspace{2cm}}$$

Are the resulting values of I_{C_Q} and V_{CE_Q} relatively close to the specified values? Calculate the percent deviations between the present values and their specified values.

If not, try to explain why. How would you correct the situation?

f. The Résumé of Theory introduced the ratio $R_1 || R_2 / \beta R_E$ as an indication of the relative stability of the design. Determine the ratio and record below.

$$R_1 || R_2 / \beta R_E \text{ (calculated)} = \underline{\hspace{2cm}}$$

g. Rebuild Fig. 11.3 with the 2N4401 transistor and repeat Part 3(e).

Exp. 11 / Procedure

V_{R_C} (measured) = _____

V_{CE_Q} (measured) = _____

I_{C_Q} (calculated from measured) = _____

β (calculated) = _____

h. Referring to the results of Part 3(**g**), how do the resulting values of I_{C_Q} and V_{CE_Q} compare to the specified values even though β has been significantly increased?

i. How do the levels of beta for this part of the experiment compare to the levels determined in Parts 1 and 2?

j. Determine the ratio of the relative stability for this configuration and compare to that calculated for the 2N3904 transistor.

$R_1 \| R_2 / \beta R_E$ (2N4401) (calculated) = _____

$R_1 \| R_2 / \beta R_E$ (2N3904) (calculated) = _____

What do the results suggest about the impact of β on the stability of the voltage-divider configuration?

k. Determine the stability factor $S(\beta) = \dfrac{\%\Delta I_C}{\%\Delta \beta}$ using the data of Parts 3(**e**) and 3(**g**).

$S(\beta)$ (calculated) = _____

Part 4. Problems and Exercises

1. a. For each configuration designed in this experiment record the resulting measured values of I_{C_Q} and V_{CE_Q} in Table 11.1.

   TABLE 11.1

Configuration	I_{C_Q}	V_{CE_Q}
Collector-feedback-bias		
Emitter-bias		
Voltage-divider-bias		

   Based on the above as compared to the specified values of $I_{C_Q} = 5$ mA and $V_{CE_Q} = 7.5$ V, do you feel satisfied with your design effort?

   Which design came closest to the specified levels?

2. Using the results of the experiment, complete Table 11.2 for the 2N4401 transistor.

   TABLE 11.2

Configuration	Stability factor	
Collector-feedback-bias	$R_B/\beta R_C =$	$S(\beta) =$
Emitter-bias	$R_B/\beta R_E =$	$S(\beta) =$
Voltage-divider-bias	$R_1 \| R_2/\beta R_E =$	$S(\beta) =$

   Is there a consistency between the stability factors of the two columns—that is, if relatively small in one column, is it relatively small in the other?

Which configuration demonstrated through $S(\beta)$ that it was the least sensitive to changes in beta? Was this expected? Why?

Which configuration has the least stability? Was this expected? Why?

The next three problems will derive the stability criteria appearing in the Résumé of Theory.

3. For the collector-feedback configuration derive an equation for the current I_C in terms of the network parameters. Then explain why the smaller the ratio $R_B/\beta R_C$ is, the less the sensitivity of I_C is to changes in beta.

4. For the emitter-bias configuration derive an equation for the current I_C in terms of the network parameters. Then explain why the smaller the ratio $R_B/\beta R_E$ is, the less the sensitivity of I_C is to changes in beta.

5. Defining the base voltage as V_B, write an equation for I_1 and I_2 of the voltage-divider configuration in terms of the network parameters. Use the resulting equations to write an equation for I_B. Substitute the fact that $V_B = V_{BE} + I_C R_E$ and rewrite the equation for I_B in a form that will support the fact that the smaller the ratio $R_1 \| R_2 / \beta R_E$ is, the less sensitive the current I_C is to changes in beta. Use the notation:

$$R_1 \| R_2 = \frac{1}{\frac{1}{R_1} + \frac{1}{R_2}}$$

Part 5. Computer Exercises

PSpice Simulation 11-1

The collector-feedback circuit shown is required to meet the following design criteria:

$$V_{CC} = 15 \text{ V}$$
$$I_{CQ} = 5 \text{ mA}$$
$$V_{CEQ} = 7.5 \text{ V}$$
Relative stability: $R_B / \beta R_C < 1.5$

$$R_B = R_{F1} + R_{F2}$$

PSpice Simulation 11-1

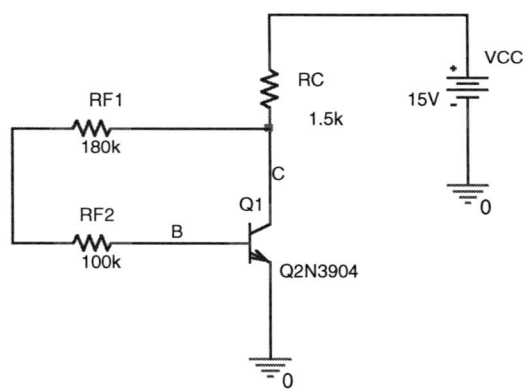

A 10% variation from the design criteria is acceptable.

1. Run a bias point analysis of this circuit to determine the circuit voltages and the collector current.

2. Calculate the β of this circuit.

3. Calculate the relative stability of this circuit.

4. Does the circuit design meet the criteria?

5. Perform a power audit of this circuit.

PSpice Simulation 11-2

After the completion of the design process, the voltage-divider circuit shown is to be bench tested. The design criteria are:

$$V_{CC} = 20 \text{ V}$$
$$I_{CQ} = 10 \text{ mA}$$
$$V_{CEQ} = 10 \text{ V}$$
$$\text{Relative stability: } (R_1 \| R_2)/(\beta R_E) < 0.07$$

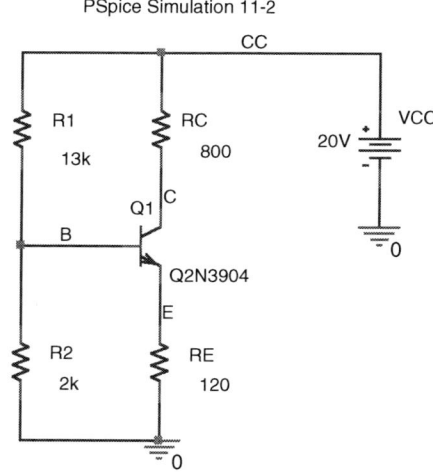

PSpice Simulation 11-2

A 10% variation from the design criteria is acceptable.

1. Run a bias point analysis of this circuit to determine the circuit voltages and the collector current.

2. Calculate the β of this circuit.

3. Calculate the relative stability of this circuit.

4. Does the circuit design meet the criteria?

5. If the criteria are not met, redesign the circuit and repeat the bias point simulation to obtain the data for the new design.

6. Does the new circuit meet the criteria?

7. If not, repeat this process until a satisfactory design is achieved.

8. Obtain a power audit of the final design.

EXPERIMENT 12

JFET Characteristics

OBJECTIVE

To obtain the output, drain, and transfer characteristics for a JFET transistor.

EQUIPMENT REQUIRED

Instruments

DMM
Curve tracer (if available)

Components

Resistors

(1) 100-Ω
(1) 1-kΩ
(1) 10-kΩ
(1) 5-kΩ potentiometer
(1) 1-MΩ potentiometer

Transistor

(1) 2N4416 (or equivalent)

Supplies

DC power supply
9 V battery with snap-on leads

EQUIPMENT ISSUED

Item	Laboratory serial no.
DMM	
DC power supply	

RÉSUMÉ OF THEORY

The junction field-effect transistor (JFET) is a unipolar conduction device. The current carriers are either electrons in an n-channel JFET or holes in a p-channel JFET. In the n-channel JFET the conduction path is an n-doped material, germanium or silicon, while in the p-channel the conduction path is p-doped germanium or silicon. Conduction through the channel is controlled by the depletion region established by oppositely doped regions in the channel. The channel is connected to two terminals, referred to as the drain and the source, respectively. For n-channel JFETs, the drain is connected to a positive voltage, and the source to a negative voltage, to establish a flow of conventional current in the channel. The polarities of the applied voltages for the p-channel JFET are opposite to those of the n-channel JFET.

A third terminal, referred to as the gate terminal, provides a mechanism for controlling the depletion region and thereby the width of the channel through which conventional flow can exist between the drain and source terminals. For an n-channel JFET, the more negative the gate-to-source voltage is, the smaller the channel width is and the less the drain-to-source current is.

This experiment will establish the relationships between the various voltages and currents flowing in a JFET. The nature of these relationships determines the range of JFET applications.

PROCEDURE

Part 1. Measurement of the Saturation Current I_{DSS} and Pinch-Off Voltage V_P

a. Construct the network of Fig. 12.1. Record the measured value of R. The 10-kΩ resistor in the input circuit is included to protect the gate circuit if the 9 V battery is applied with the wrong polarity and the potentiometer is set on its maximum value.

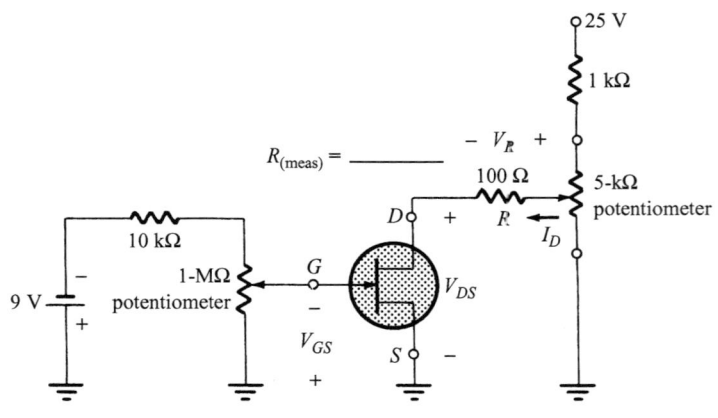

Figure 12-1

b. Vary the 1-MΩ potentiometer until $V_{GS} = 0$ V. Recall that $I_D = I_{DSS}$ when $V_{GS} = 0$ V.

c. Set V_{DS} to 8 V by varying the 5-kΩ potentiometer. Measure the voltage V_R.

V_R (measured) = _____

d. Calculate the saturation current from $I_{DSS} = I_D = V_R/R$ using the measured resistor value and record below.

I_{DSS} (calculated from measured) = _____

e. Maintain V_{DS} at about 8 V and reduce V_{GS} until V_R drops to 1 mV. At this level $I_D = V_R/R = 1$ mV/100 Ω = 10 μA ≅ 0 mA. Recall that V_P is the voltage V_{GS} that results in $I_D = 0$ mA. Record the pinch-off voltage below:

V_P (measured) = _____

f. Check with two other groups in your laboratory area and record their levels of I_{DSS} and V_P.

1. I_{DSS} = _____ , V_P = _____
2. I_{DSS} = _____ , V_P = _____

Based on the above, are I_{DSS} and V_P the same for all 2N4416 transistors?

g. Using the calculated and measured values of I_{DSS} and V_P, sketch the transfer characteristics for the device in Fig. 12.2 using Shockley's equation. Plot at least 5 points on the curve.

$$I_D = I_{DSS}\left(1 - \frac{V_{GS}}{V_P}\right)^2 \tag{12.1}$$

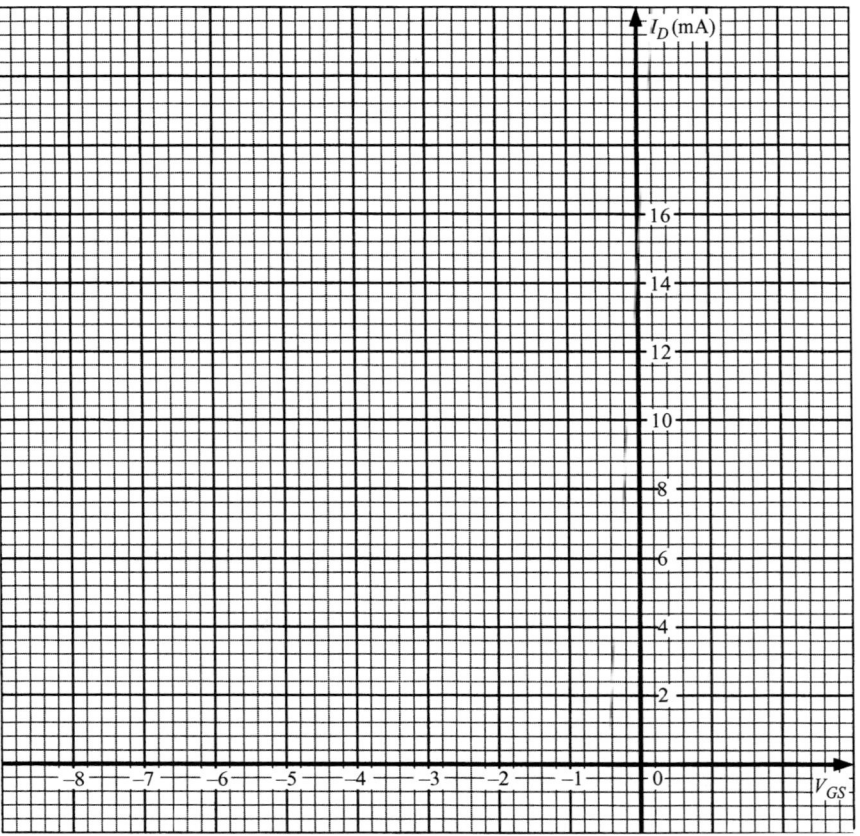

Figure 12-2 Transfer characteristics: 2N4416

Part 2. Output Characteristics

This part of the experiment will determine the I_D versus V_{DS} characteristics for an n-channel JFET.

- **a.** Using the network of Fig. 12.1, vary the two potentiometers until $V_{GS} = 0$ V and $V_{DS} = 0$ V. Determine I_D from $I_D = V_R/R$ using the measured value of R and record in Table 12.1.
- **b.** Maintain V_{GS} at 0 V and increase V_{DS} through 14 V (in 1 volt steps) and record the calculated value of I_D. Be sure to use the measured value of the 100-Ω resistance in your calculations.
- **c.** Vary the 1-MΩ potentiometer until $V_{GS} = -1$ V. Maintaining V_{GS} at this level, vary V_{DS} through the levels of Table 12.1 and record the calculated value of I_D.
- **d.** Repeat Part 2(**c**) for the values of V_{GS} appearing in Table 12.1. Discontinue the process once V_{GS} exceeds V_P.
- **e.** Plot the output characteristics for the JFET on the graph of Fig. 12.3.
- **f.** Does the plot verify the conclusions of Part 1? That is, is the average value of I_D for $V_{GS} = 0$ V relatively close to I_{DSS}? Is the value of V_{GS} that results in $I_D = 0$ mA close to V_P?

TABLE 12.1

V_{GS} (V)	0	−1.0	−2.0	−3.0	−4.0	−5.0	−6.0
V_{DS} (V)	I_D (mA)	I_D (mA)	I_D (mA)	I_D (mA)	I_D (mA)	I_D (mA)	I_D (mA)
0.0							
1.0							
2.0							
3.0							
4.0							
5.0							
6.0							
7.0							
8.0							
9.0							
10.0							
11.0							
12.0							
13.0							
14.0							

I_{DSS} (Fig. 12.3) = _____
I_{DSS} (Part 1) = _____
V_P (Fig. 12.3) = _____
V_P (Part 1) = _____

Part 3. Transfer Characteristics

This part of the experiment will determine the I_D vs. V_{GS} transfer characteristics frequently used in the analysis of JFET networks. Ideally, the transfer characteristics as determined by Shockley's equation assume that the effect of V_{DS} can be ignored and the characteristic curves of Fig. 11.3 for a given V_{GS} are considered horizontal. The following will show that the transfer curve does vary slightly with V_{DS} but not to the point where concern should develop about using Shockley's equation.

For this part of the experiment all the data can be obtained from Table 12.1. There is no experimental work in this part.

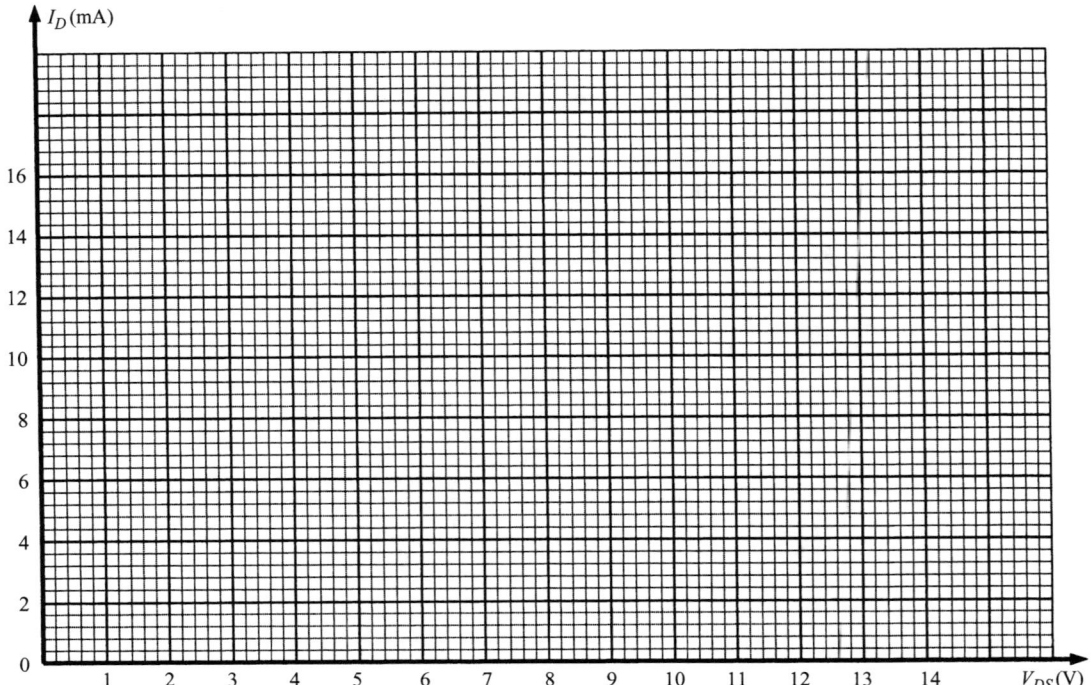

Figure 12-3 Drain-current curve: 2N4416.

a. At $V_{DS} = 3$ V record the values of I_D for the range of V_{GS} in Table 12.2 using the data of Table 12.1.

TABLE 12.2

V_{DS}	3 V	6 V	9 V	12 V
V_{GS}	I_D (mA)	I_D (mA)	I_D (mA)	I_D (mA)
0 V				
−1 V				
−2 V				
−3 V				
−4 V				
−5 V				
−6 V				

b. Repeat Part 3(a) for $V_{DS} = 6$ V, 9 V, and 12 V.
c. For each level of V_{DS} plot I_D vs. V_{GS} on the graph of Fig. 12.4. Plot each curve carefully and label each curve with the value of V_{DS}.
d. Is it reasonable (on an approximate basis) to assume the family of curves of Fig. 12.4 can be replaced by a single curve defined by Shockley's equation?

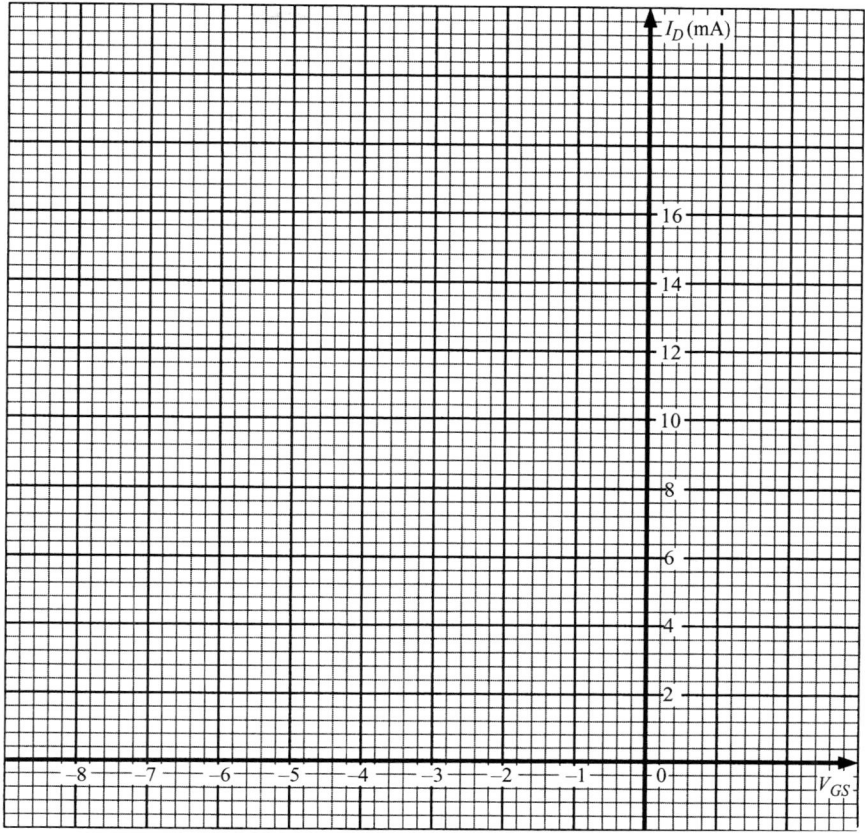

Figure 12-4 Pinch-off voltage curve: 2N4416.

Part 4. Determination of the JFET Characteristics Using a Commercial Curve Tracer

a. If available, use a curve tracer to obtain an output set (I_D vs. V_{DS}) of characteristics for the 2N4416 JFET.

b. Reproduce the characteristics on the graph of Fig. 12.5.

c. Compare the characteristics to those obtained in Part 2, Fig. 12.3. Note that the scales are the same to permit a direct comparison.

d. From your data obtained in Fig. 12.5 draw the transfer characteristics in Fig. 12.6. Compare this graph with Fig. 12.4 in Part 3. Use as many data points from Fig. 12.5 as you feel are required to obtain the desired curves.

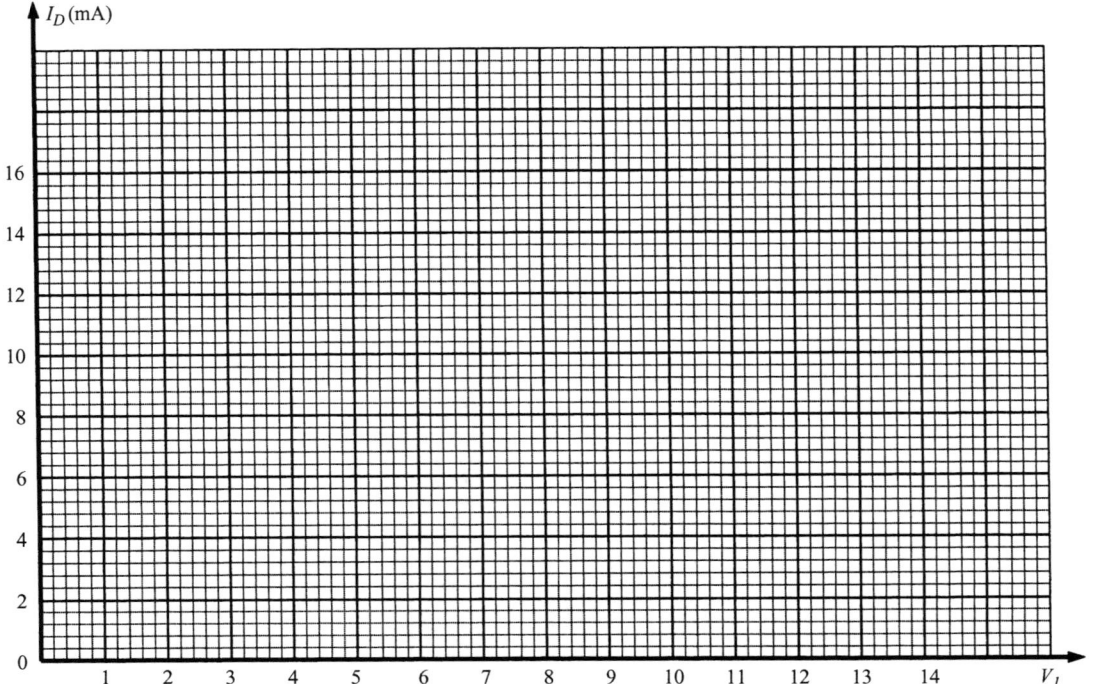

Figure 12-5 Drain-source characteristic: 2N4416.

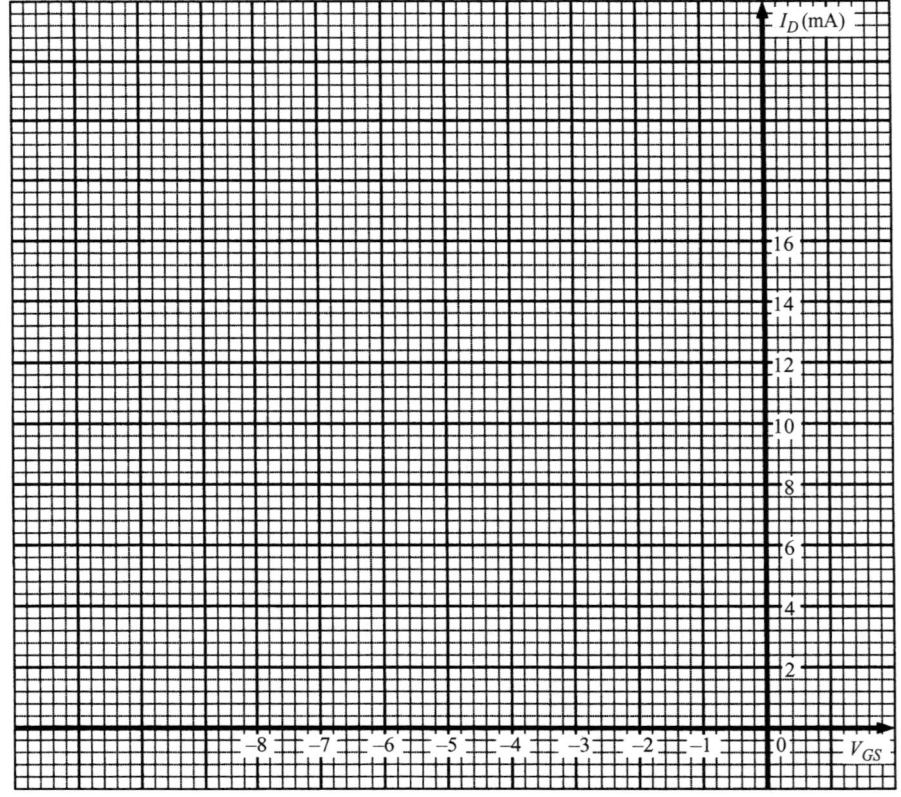

Figure 12-6 Transfer characteristic: 2N4416.

Part 5. Problems and Exercises

1. Given I_D and V_{GS} at a particular point on Shockley's curve, can the values of I_{DSS} and V_P be determined? If so, how? If not, why not?

2. **a.** Write Shockley's equation in a form that will provide V_{GS} in terms of I_{DSS}, V_P, and I_D.

   b. Given $I_{DSS} = 10$ mA, $V_P = -5$ V, and $I_D = 4$ mA, find the value of V_{GS}.

   V_{GS} (calculated) = _____

3. The transconductance, g_m, of a JFET is an important quantity in the AC analysis of JFET amplifiers. Its magnitude is defined by the slope of Shockley's equation at a point on the characteristics. The application of calculus techniques to Shockley's equation will result in the following equation for g_m:

$$g_m = g_{mo}\left(1 - \frac{V_{GS}}{V_P}\right) \quad (12.1)$$

with

$$g_{mo} = \frac{2\,I_{DSS}}{|V_P|} \quad (12.2)$$

which is the transconductance at $V_{GS} = 0$ V.

   a. Using the experimental results of Part 1, determine g_{mo}.

   g_{mo} (calculated) = _____

b. Referring to the transfer curve of Fig. 12.2, is the slope of Shockley's equation a maximum at $V_{GS} = 0$ V?

Based on the above, can we assume that g_{mo} calculated in Exercise 3(**a**) is the maximum value of g_m?

c. Determine g_m at $V_{GS} = V_P$.

g_m (calculated) = _____

Referring to the transfer curve of Fig. 12.2, is the slope of Shockley's equation a minimum at $V_{GS} = V_P$? In fact, what slope would you expect it to have exactly at $V_{GS} = V_P$?

d. Determine g_m at $V_{GS} = \frac{1}{4}V_P$, $\frac{1}{2}V_P$, and $\frac{3}{4}V_P$, and plot the curve of g_m on Fig. 12.7.

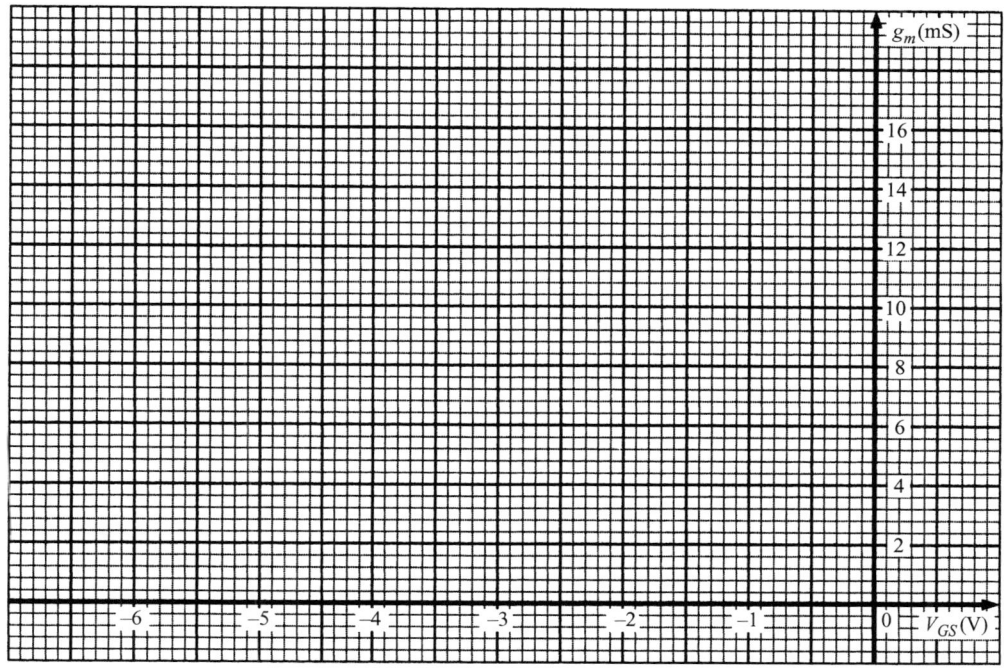

Figure 12-7 Transconductance versus V_{GS} of 2N4416.

    e. Referring to the transfer curve of Fig. 12.2, does the slope increase with less negative values of V_{GS}? Is your conclusion verified by the plot of Fig. 12.7?

    f. What does the fact that the graph of Fig. 12.7 is a straight line tell you about the curve resulting from Shockley's equation?

Part 6. Computer Exercises

PSpice Simulation 12-1

This circuit corresponds closely to that of Figure 12-1. The difference is the JFET used.

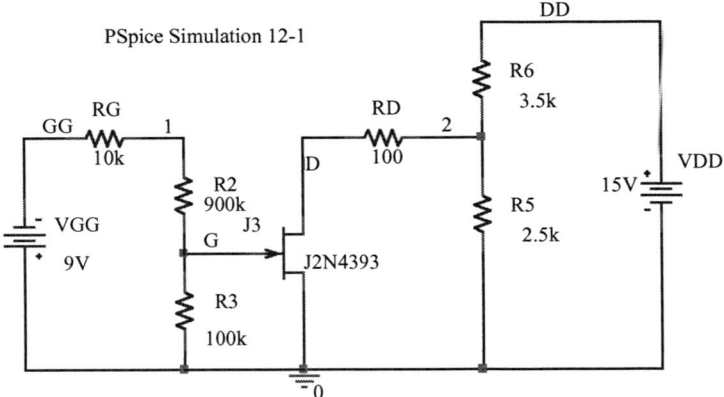

For the indicated operating point of this circuit, run a bias point simulation and obtain:

1. The drain-to-source voltage.

2. The gate-to-source voltage.

3. The drain current.

4. The gate current.

PSpice Simulation 12-2

Determine the drain characteristics for the J2N4393 JFET using the circuit shown.

PSpice Simulation 12-2: Drain characteristics

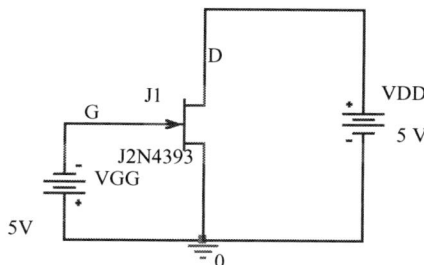

To do this:

1. Run a DC Sweep simulation.

2. For the primary sweep of VDD, specify a sweep range from 0.1 V to 5 V in steps of 0.1 V.

3. For the secondary sweep (nested sweep) of VGG, specify a sweep range from 0.1 V to 5 V in steps of 0.4 V. The increment may be changed in size to suit view.

4. On the Probe plot, with the x-axis in units of V_VDD, or V(D), obtain a trace of the drain current ID(J1).

5. What is the value of the saturation current IDSS?

6. What is the value of the pinch-off voltage?

Determine the transfer characteristic of the J2N4393 JFET.

To do this:

1. Run a DC Sweep simulation.

2. Keep VDD constant at 5 volts.

3. Primary sweep the voltage source VGG from –3 V to 0 V in steps of 0.1 V.

4. On the Probe plot, with the x-axis in units of V_VGG, or V(G), obtain a trace of the drain current ID(J1).

5. Compare the values of the saturation current IDSS and the pinch-off voltage with those attained from the drain characteristics.

Name _____
Date _____
Instructor _____

EXPERIMENT 13

JFET Bias Circuits

OBJECTIVE

To analyze fixed-, self-, and voltage-divider-bias JFET networks.

EQUIPMENT REQUIRED

Instrument

DMM

Components

Resistors

(1) 1-kΩ
(1) 1.2-kΩ
(1) 2.2-kΩ
(1) 3-kΩ
(1) 10-kΩ
(1) 10-MΩ
(1) 1-kΩ potentiometer

Transistors

(1) JFET 2N4166 (or equivalent)

Supplies

DC power supply
9 V battery with snap-on leads

EQUIPMENT ISSUED

Item	Laboratory serial no.
DMM	
DC power supply	

RÉSUMÉ OF THEORY

In this experiment, three different biasing circuits will be analyzed. In theory, the procedure for biasing a JFET is the same as that for a BJT. In particular, given the drain curve characteristics of the JFET and the external circuit connected to the JFET, a load line is constructed involving V_{DD}, V_{DS}, and I_D. The intersection of that load line with the drain curve characteristics determines the quiescent operating point for the JFET. The characteristics of the device are a property of the JFET; by contrast, the load line is dependent on the external circuit elements connected to the JFET. The quiescent operating point is determined by the intersection of the two curves.

In practice, JFETs, even of the same type, show considerable variation in their drain curve characteristics. As a result, manufacturers often do not publish these curves; rather, the values for the saturation current and the pinch-off voltage are given as part of the specifications. This leads to an alternative approach to determine the quiescent condition for a JFET.

To begin, the transconductance curve, which shows the relationship between V_{GS} and I_D for a particular JFET, is constructed from the saturation current, the pinch-off voltage, and Shockley's equation. Next, a bias curve is constructed sensitive to the external circuit elements connected to the JFET. The quiescent condition is determined by the intersection of the two curves.

PROCEDURE

Part 1. Fixed-Bias Network

For the fixed-bias configuration, V_{GS} will be set by an independent DC supply. The vertical lines of constant V_{GS} will intersect the transfer curve developed from Shockley's equation.

a. Construct the network of Fig. 13.1. Record the measured value of R_D.

b. Set V_{GS} to zero volts and measure the voltage V_{R_D}. Calculate I_D from $I_D = V_{R_D}/R_D$ using the measured value of R_D. Since $V_{GS} = 0$ V the resulting drain current is the saturation value I_{DSS}. Record below.

I_{DSS} (from measured) = _____

c. Make V_{GS} increasingly negative until $V_{R_D} = 1$ mV (and $I_D = V_{R_D}/R_D \cong 1$ µA). Since I_D is very small ($I_D \cong 0$ A), the resulting value of V_{GS} is the pinch-off voltage V_P. Record below.

V_P (measured) = _____

These values will be used throughout the experiment.

d. Using the values above for I_{DSS} and V_P, sketch the transfer curve on Fig. 13.2 using Shockley's equation.

e. If $V_{GS} = -1$ V, determine I_{D_Q} from the curve of Fig. 13.2. Show all work in Fig. 13.2. Label the straight line defined by V_{G_S} as the fixed-bias line.

I_{D_Q} (calculated) = _____

f. Set $V_{GS} = -1$ V in Fig. 13.1 and measure V_{R_D}. Calculate I_{D_Q} using the measured value of R_D and record below. This is the measured value of I_D.

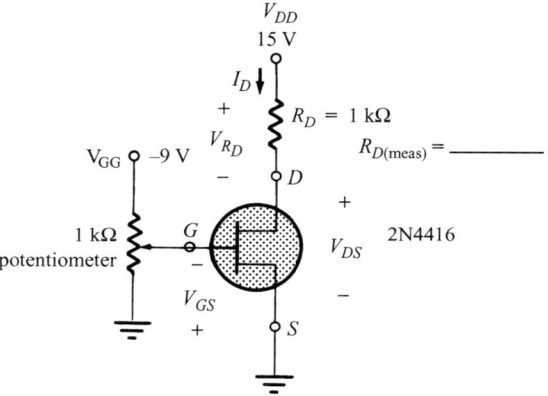

Figure 13-1 Fixed-bias circuit.

V_{R_D} (measured) = _____

I_{D_Q} (from measured) = _____

g. Compare the measured and calculated values of I_{D_Q}.

Part 2. Self-Bias Network

In the self-bias configuration, the magnitude of V_{GS} is defined by the product of the drain current I_D and source resistance R_S. The network bias line will start at the origin and intersect the transfer curve at the quiescent (DC) point of operation. The resulting drain current and gate-to-source voltage can then be determined from the graph by drawing a horizontal and a vertical line from the quiescent point to the axis, respectively. *Note:* The larger the

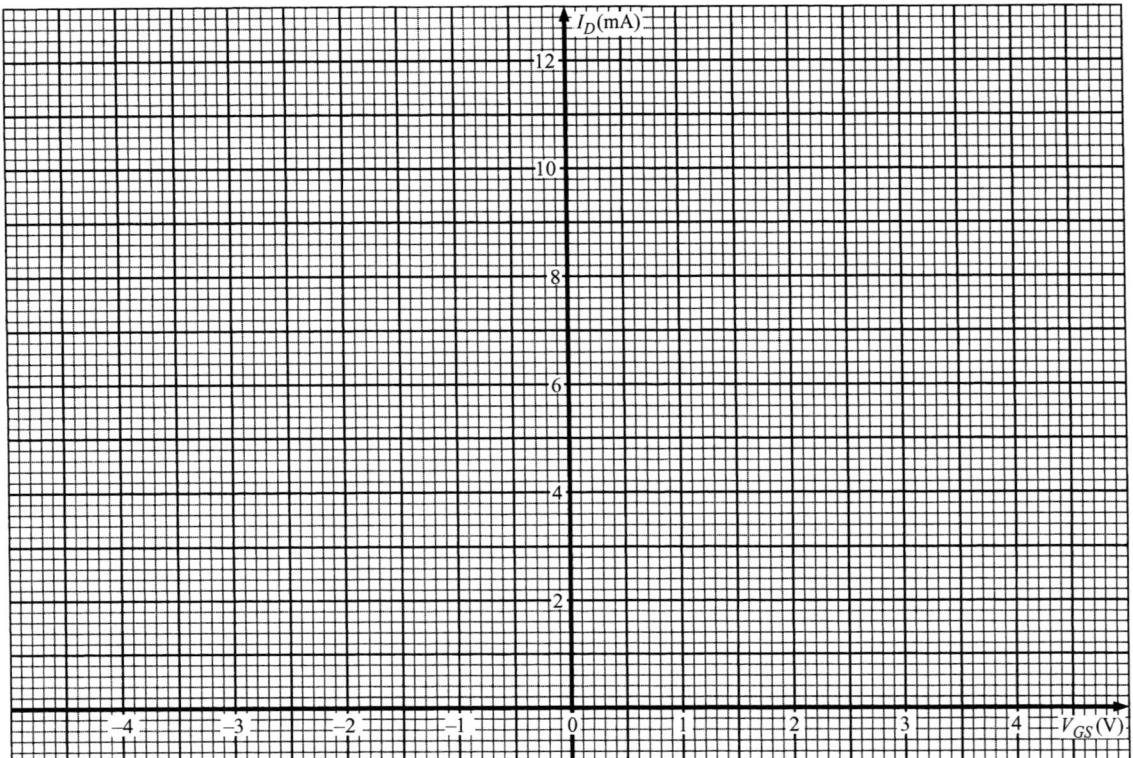

Figure 13-2 Bias lines and transfer characteristics.

source resistance, the more horizontal the bias line and the less the resulting drain current.

    a. Construct the network of Fig. 13.3. Record the measured value of R_D and R_S.

    b. Draw the self-bias line defined by $V_{GS} = -I_D R_S$ in Fig. 13.2 and find the network Q point. Record the quiescent values of I_{D_Q} and V_{GS_Q} below. Label the straight line as the self-bias line.

$$I_{D_Q} \text{ (calculated)} = \underline{\hspace{2cm}}$$
$$V_{GS_Q} \text{ (calculated)} = \underline{\hspace{2cm}}$$

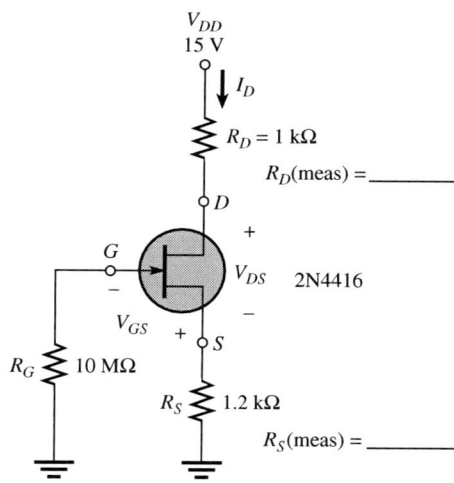

Figure 13-3 Self-bias circuit.

c. Calculate the values of V_{GS}, V_D, V_{DS}, and V_G and record below.

V_{GS} (calculated) = _____
V_D (calculated) = _____
V_S (calculated) = _____
V_{DS} (calculated) = _____
V_G (calculated) = _____

d. Measure the voltages V_G, V_{DS}, V_D, V_S, and V_G and compare with the results above using the equation

$$\% \text{ difference} = \frac{|V_{meas} - V_{calc}|}{|V_{calc}|} \times 100\% \qquad (13.1)$$

V_{GS} (measured) = _____
V_D (measured) = _____
V_S (measured) = _____
V_{DS} (measured) = _____
V_G (measured) = _____

% (V_{GS}) (calculated) = _____
% (V_D) (calculated) = _____
% (V_S) (calculated) = _____
% (V_{DS}) (calculated) = _____
% (V_G) (calculated) = _____

Part 3. Voltage-Divider-Bias Network

In the voltage-divider-bias configuration, V_{GS} is determined by a voltage-divider-bias voltage and voltage drop across the source resistance. That is, for the network of Fig. 13.4,

$$V_G = \frac{R_2 V_{DD}}{R_1 + R_2}$$

and

$$V_{GS} = V_G - I_D R_S$$

a. Construct the network of Fig. 13.4. Record the measured resistor values.

b. Using the I_{DSS} and V_P determined in Part 1, draw the voltage-divider-bias line in Fig. 13.2 and find the network Q point. Label the resulting straight line as the voltage-divider line.

To draw the bias line, determine two points as follows and then connect the two points with a straight line.

For $V_{GS} = V_G - I_D R_S$
If $I_D = 0$ mA   then $V_{GS} = V_G - (0)(R_S) = V_G$

and if $V_{GS} = 0$ V   then $I_D = \dfrac{V_G}{R_S}$

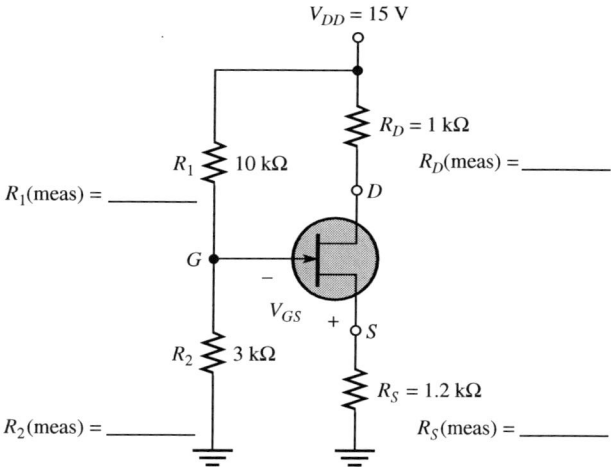

Figure 13-4 Voltage-divider-bias circuit.

c. Draw a straight line through the above two points and extend it until it intersects the transfer curve. The coordinates of that intersection determine the quiescent values of I_D and V_{GS}. Record them below.

I_{D_Q} (calculated) = _____

V_{GS_Q} (calculated) = _____

d. Calculate the theoretical values of V_D, V_S, and V_{DS} and record below.

V_D (calculated) = _____
V_S (calculated) = _____
V_{DS} (calculated) = _____

e. Measure the voltages V_{GS_Q}, V_D, V_S, and V_{DS} and record below.

V_{GS_Q} (measured) = _____
V_D (measured) = _____
V_S (measured) = _____
V_{DS} (measured) = _____

f. Calculate the percent difference between the measured and calculated values using Eq. 13.1.

% (V_{GS_Q}) (calculated) = _____
% (V_D) (calculated) = _____
% (V_S) (calculated) = _____
% (V_{DS}) (calculated) = _____

g. Calculate I_{D_Q} from the measured voltages of Part 3(e) and compare to the value determined in Part 3(c). I_{D_Q} can be found using

$$I_{D_Q} = \frac{V_{DD} - V_D}{R_D}$$

and the measured values of V_D and R_D. Record below and calculate the % difference.

Exp. 13 / Part 4. ComputerExercises

I_{D_Q} (measured) = _____

% (I_{D_Q}) (calculated) = _____

Part 4. Computer Exercises

PSpice Simulation 13-1

Perform a bias point simulation for the self-bias circuit shown. Perform the following steps:

1. Obtain the drain current.

2. Obtain the drain-to-source voltage.

3. Obtain the gate-to-source voltage.

4. Obtain the DC power delivered by the source VDD.

5. Obtain the DC power absorbed by each of the resistors.

6. Explain the value of the DC power to the resistor RG.

PSpice Simulation 13-1: Self-bias circuit

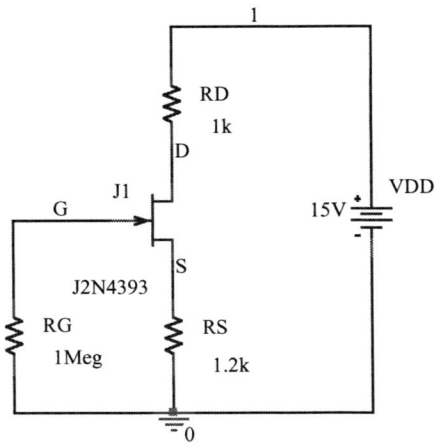

7. Obtain the power absorbed by the JFET.

8. If a drain-to-source voltage of ½(VDD) with no more than a 10% positive or negative variation is desired, does this circuit meet the criteria?

PSpice Simulation 13-2

Perform a bias point simulation for the voltage-divider-bias circuit shown. Perform the following steps:

1. Obtain the drain current.

2. Obtain the drain-to-source voltage.

3. Obtain the voltage across resistor RD.

Exp. 13 / Part 4. Computer Exercises

4. Obtain the gate-to-source voltage.

5. Obtain the source voltage.

6. Obtain the DC power delivered by the source VDD.

7. Obtain the DC power absorbed by each of the resistors.

8. Obtain the power absorbed by the JFET.

9. If a drain-to-source voltage of ½(VDD) with no more than a 10% positive or negative variation is desired, does this circuit meet the criteria?

PSpice Simulation 13-2: Voltage-divider-bias circuit.

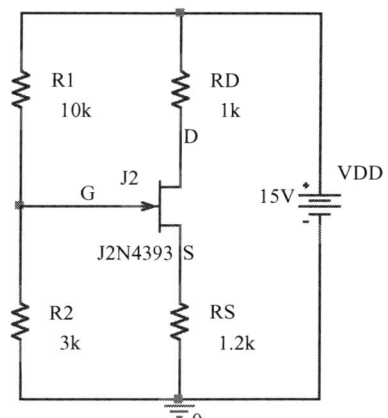

Part 5. Problems and Exercises

1. a. Record your values of I_{DSS} and V_P and the values for both quantities from two other lab groups.

    I_{DSS} (your lab group) = _____, V_P = _____
    I_{DSS} (other lab groups) = _____, V_P = _____
    I_{DSS} = _____, V_P = _____

    b. Is the range more than you expected for a specified JFET? What effect will the range of values have on the design process?

2. For the self-bias configuration what is the effect of increasing values of R_S on the resulting Q-point? That is, does an increasing value of R_S result in an increase or decrease in the level of I_{D_Q}? What is the effect of an increasing value of R_S on V_{GS_Q}? Explain in some detail.

3. What value of source resistance (R_S) would make the quiescent drain current equal to 1/2 the saturation level (I_{DSS}) for the self-bias configuration? Use the parameter values of Fig. 13.3 and your level of I_{DSS} and V_P.

    R_S (calculated) = _____

4. What value of source resistance (R_S) would make the quiescent drain current I_{D_Q} equal to 1/2 the saturation level (I_{DSS}) for the voltage-divider configuration of Fig. 13.4? Use the parameter values of Fig. 13.4 and your level of I_{DSS} and V_P.

R_S (calculated) = _____

EXPERIMENT 14

Design of JFET Bias Circuits

OBJECTIVES

1. To design a self-bias JFET circuit for specified bias conditions.
2. To design a voltage-divider-bias JFET circuit for specified bias conditions.
3. To test both of these circuits and, if necessary, redesign them.

EQUIPMENT REQUIRED

Instruments

Dual-trace oscilloscope
DMM
DC power supply
9 V battery with snap-on leads

Components

Resistors

(1) 1-kΩ
(1) 1-kΩ potentiometer

Since this is a design experiment, a number of resistor values will have to be determined and requested from the stockroom.

Transistors

(1) 2N4416 JFET or equivalent

Exp. 14 / Design of JFET Bias Circuits

EQUIPMENT ISSUED

Item	Laboratory serial no.
Oscilloscope	
DMM	
DC power supply	
Signal generator	

RÉSUMÉ OF THEORY

Like the BJT, the junction field-effect transistor can operate in three modes: cutoff, saturation, and linear. The physical characteristics of the JFET and the external circuit elements connected to it determine the mode of operation. In this experiment, the JFET is biased in the linear mode in accordance with a given set of circuit specifications.

JFETs, even of the same type, show a considerable variation in their characteristics. In consequence, manufacturers rarely publish the drain characteristics of JFETs but specify both the saturation current I_{DSS} and the pinch-off voltage V_P. The designer can construct the transfer characteristic from these two values and any intermittent values of I_D and V_{GS} from Shockley's equation.

The transfer curves together with the operating specifications will be used to determine needed values for the various circuit elements to be used in the two bias designs. In both cases, a procedure will be suggested that will place the Q-point at or near the specified DC operating conditions and will allow for a specified AC signal to be amplified without distortion. Upon completion of the design, the experimenter will construct the actual circuit and take the DC measurements to ensure correct operation of the circuit. In case the design does not meet the specifications, a corrective procedure is suggested.

PROCEDURE

Part 1. Determining I_{DSS} and V_P

This part of the experiment will determine I_{DSS} and V_P for the JFET to be employed in the design process of this experiment.

    a. Construct the network of Fig. 14.1. Insert the measured value of R_D.

    b. Set V_{GS} to zero volts and measure V_{R_D}. Calculate $I_D = I_{DSS} = V_{R_D}/R_D$ using the measured resistance value and record below.

I_{DSS} (measured) = _____

    c. Make V_{GS} increasingly negative until V_{R_D} = 1 mV (and $I_D = V_{R_D}/R_D \cong 1\ \mu A$). Since I_D is small ($I_D \cong 0$ A), the resulting value of V_{GS} is the pinch-off voltage V_P. Record below.

V_P (measured) = _____

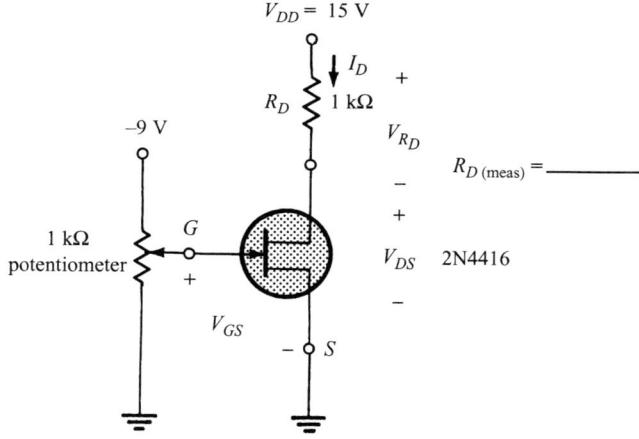

Figure 14-1 Network to determine I_{DSS} and V_P.

Part 2. Self-Bias Circuit Design

This part of the experiment will determine R_D and R_S for the self-bias configuration of Fig. 14.2. The Q-point is calculated at $I_{D_Q} = \frac{1}{2}(I_{DSS})$, $V_{DS_Q} = \frac{1}{2}(V_{DS_{max}})$ with $V_{DD} = 2V_{DS_Q}$.

a. Using the specified Q-point and the results of Part 1, with the constraint that $V_{DS_{max}} = 30$ V, calculate I_{D_Q}, V_{DS_Q}, and V_{DD}.

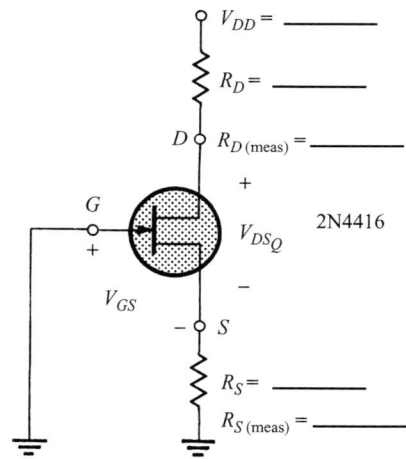

Figure 14-2 Self-bias configuration to be designed.

I_{D_Q} (calculated) = _____

V_{DS_Q} (calculated) = _____

V_{DD} (calculated) = _____

b. Using the I_{DSS} and V_P determined in Part 1, sketch the transfer characteristics in Fig. 14.3.

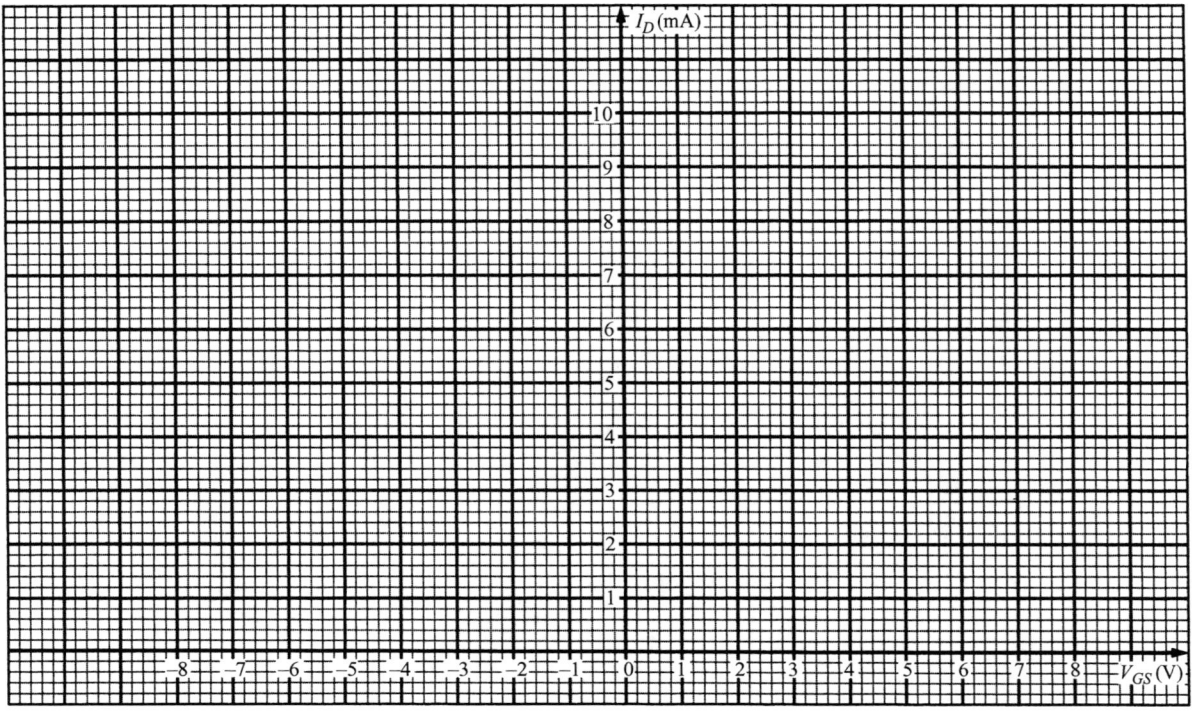

Figure 14-3 Transfer characteristic for the 2N4416 JFET.

c. The choice of $I_{D_Q} = \frac{1}{2}(I_{DSS})$ will permit a maximum swing in collector current for the AC domain. Draw a horizontal line from $\frac{1}{2}(I_{DSS})$ on the I_C axis to the transfer curve of Fig. 14.3 and label the intersection as the Q-point to define the self-bias line. Label the resulting line as the self-bias line for future reference.

d. Calculate the value of R_S from

$$R_S = \left| \frac{\Delta V}{\Delta I} \right| = \left| \frac{\Delta V_{GS}}{\Delta I_D} \right|$$

where | | specifies the magnitude of the quantity and Δ the change in each quantity from the origin to the Q-point.

R_S (calculated) = _____

Determine the closest commercial value (available in the laboratory) to the calculated R_S and insert its resistance value in the space provided on Fig. 14.2. In addition, record the calculated value for V_{DD} from Part 2(**a**).

R_S (commercial value) = _____

Insert the commercial and measured value of R_S in the space provided on Fig. 14.2.

e. The resistance of R_D will now be determined by an application of Kirchhoff's voltage law to the output circuit of Fig. 14.2 followed by the use of Ohm's law. For the output circuit of Fig. 14.2,

$$V_{R_D} = V_{DD} - V_{DS_Q} - V_{R_S}$$

Determine $V_{R_S} = I_{D_Q} R_S$ (where R_S is the measured resistance value) and substitute the levels of V_{DS_Q} and V_{DD} from Part 2(**a**) to determine V_{R_D}.

V_{R_D} (calculated) = _____

Determine R_D using I_{D_Q} from Part 2(**a**) and Ohm's law.

R_D (calculated) = _____

Determine the closest commercial resistance value (available in the laboratory) to R_D and record below.

R_D (commercial value) = _____

Insert the commercial value and the measured value of R_D in the space provided in Fig. 14.2.

f. Using the commercial values of R_D and R_S and the calculated value of V_{DD}, construct the network of Fig. 14.2. Energize the network and measure V_{DS_Q} and V_{R_D}. Use the measured value of R_D to calculate I_{D_Q}. Record the levels of V_{DS_Q} and I_{D_Q} below.

V_{DS_Q} (measured) = _____

I_{D_Q} (measured) = _____

Record the design levels of I_{D_Q} and V_{DS_Q} calculated in Part 2(**a**).

V_{DS_Q} (calculated) = _____

I_{D_Q} (calculated) = _____

g. Compare your data from Part 2(**a**) with the data from Part 2(**f**).

Perform a percent deviation calculation. Use the data from Part 2(**a**) as the standard.

How could you reduce the size of that deviation if such was needed?

h. Borrow the 2N4416 JFET from the group next to you and insert it in your network using the values of R_D, R_S, and V_{DD} determined above. Measure the resulting levels of V_{DS_Q} and V_{R_D} and calculate I_{D_Q} as above. Record the results below.

V_{DS_Q} (measured) = _____

I_{D_Q} (calculated) = _____

In addition, record the levels of I_{DSS} and V_P for the borrowed JFET transistor.

I_{DSS} (borrowed JFET) = _____

V_P (borrowed JFET) = _____

How do the values of V_{DS_Q} and I_{D_Q} compare with the specified design values of Part 2(**a**) for this borrowed transistor? Use the recorded values of I_{DSS} and V_P for the other JFET to help explain any variations.

What do the results obtained above tell you about the design process when using JFETs of the same nameplate data?

The specification sheet for the 2N4416 JFET specifies a range of 5 mA to 15 mA for I_{DSS} and -1 V to -6 V for V_P. Are the levels of I_{DSS} and V_P obtained for your JFET and the borrowed JFET in this range? Are the average values of 10 mA and -3.5 V good choices for a first design effort if I_{DSS} and V_P are unknown quantities?

Part 3. Voltage-Divider Circuit Design

This part of the experiment will determine the value of R_D, R_S, R_1, and R_2 for the voltage-divider configuration of Fig. 14.4. The Q-point is to be established at

$$I_{D_Q} = 4 \text{ mA}$$
$$\text{and} \quad V_{DS_Q} = 8 \text{ V}$$

Additional specifications:

$$V_{DD} = 20 \text{ V}$$
$$R_2 = 10 R_S$$

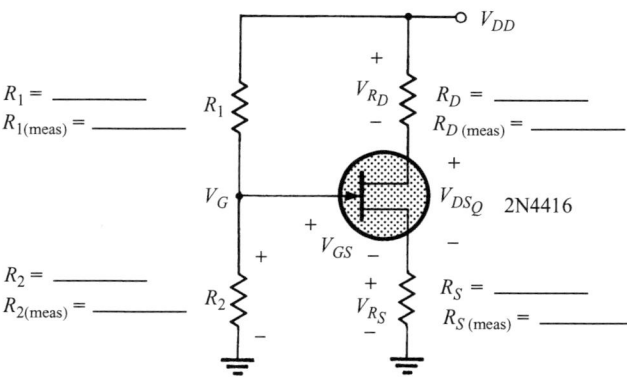

Figure 14-4 Voltage-divider configuration to be designed.

a. First locate the Q-point on the transfer curve of Fig. 14.3 by drawing a horizontal line from $I_{D_Q} = 4$ mA on the vertical axis to the transfer curve. At the Q-point draw a vertical straight line to determine the corresponding value of V_{GS}.

V_{GS} (calculated) = _____

The network equation for V_{GS} is

$$V_{GS} = V_G - I_D R_S \qquad (14.1)$$

b. A design decision must now be made. Both V_G and R_S are undefined quantities. However, we know V_G must be a positive voltage, and too small a level of voltage will result in a very low resistance level for R_S. A low level of R_S will in turn result in an undesirably high level of DC dissipation during operation. The upper voltage of V_G is limited by the available supply. As a compromise, let us choose V_G to be equal to the magnitude of the pinch-off voltage V_P. That is,

$$V_G = |V_P| \qquad (14.2)$$

The level of R_S can then be determined using Eq. 14.1.

Using the specified quiescent conditions, the resulting level of V_{GS}, and the V_P for your 2N4416 transistor, calculate the required level of R_S.

R_S (calculated) = _____

Determine the closest commercial value (available in the laboratory) to R_S and insert below.

R_S (commercial value) = _____

Record the commercial and calculated values for R_S in Fig. 14.4.

Using the measured value of R_S, V_{DS_Q}, and I_{D_Q}, determine the resulting value of V_G using Eq. 14.1. Is it relatively close to the value determined by V_P?

V_G (calculated) = _____

c. For the output circuit

$$V_{R_D} = V_{DD} - V_{DS_Q} - V_{R_S}$$

Calculate V_{R_S} from $V_{R_S} = I_{D_Q} R_S$ (using the measured resistance value) and substitute the specified values of V_{DD} and V_{DS_Q} to determine V_{R_D}.

V_{R_D} (calculated) = _____

Determine R_D using I_{D_Q} and V_{R_D}.

R_D (calculated) = _____

Determine the closest commercial value (available in the laboratory) to R_D and record below.

R_D (commercial value) = _____

Record the commercial and measured values of R_D in Fig. 14.4.

d. R_1 and R_2 will now be determined using the following equation:

$$V_G = \frac{R_2 V_{DD}}{R_1 + R_2} \tag{14.3}$$

and the specification

$$R_2 = 10 R_S \tag{14.4}$$

Using the commercial value of R_S, calculate the level of R_2 using Eq. 14.4.

R_2 (calculated) = _____

Determine the closest commercial value (available in the laboratory) to R_2.

R_2 (calculated) = _____

Record the commercial and measured values for R_2 in Fig. 14.4.

Using the specified value of V_{DD}, the calculated value of V_G (from the measured value of R_S), and the commercial value of R_2, calculate the value of R_1 using Eq. 14.3.

R_1 (calculated) = _____

Determine the closest commercial value (available in the laboratory) to R_1 and record below.

R_1 (commercial value) = _____

Record the commercial and measured values of R_1 in Fig. 14.4.

e. Using the commercial values of R_D, R_S, R_1, and R_2 and the specified value of V_{DD}, construct the network of Fig. 14.4. Energize the network and measure V_{DS_Q} and V_{R_D}. Using the measured value of R_D, calculate I_{D_Q}. Record the levels of V_{DS_Q} and I_{D_Q} below.

V_{DS_Q} (measured) = _____

I_{D_Q} (calculated) = _____

Record the specified design values for V_{DS_Q} and I_{D_Q} below.

V_{DS_Q} (specified) = _____

I_{D_Q} (specified) = _____

Exp. 14 / Procedure

f. Using the following equations, determine the percent difference between the specified and measured values of I_{D_Q} and V_{DS_Q}.

$$\% I_{D_Q} = \frac{|I_{D_{Q(\text{specified})}} - I_{D_{Q(\text{calculated})}}|}{|I_{D_{Q(\text{specified})}}|} \quad (14.5)$$

$$\% V_{DS_Q} = \frac{|V_{DS_{Q(\text{specified})}} - V_{DS_{Q(\text{measured})}}|}{|V_{DS_{Q(\text{specified})}}|} \quad (14.6)$$

$\% I_{D_Q}$ (calculated) = _____

$\% V_{DS_Q}$ (calculated) = _____

Are you satisfied with your design effort? Be specific.

g. If the percent difference is more than 10%, how would you improve the design? Take careful note of your voltage-divider bias line on Fig. 14.3 when you consider alternative designs.

Make adjustments in your design to reduce the percent differences in I_{D_Q} and V_{DS_Q} to less than 10%. Record the new values of R_D, R_S, R_1, and R_2 below. In addition, record the new values of I_{D_Q} and V_{DS_Q} and calculate the percent differences with the specified levels.

R_D (commercial value) = _____

R_S (commercial value) = _____

R_1 (commercial value) = _____

R_2 (commercial value) = _____

I_{D_Q} (calculated) = _____

V_{DS_Q} (measured) = _____

% I_{D_Q} (calculated) = _____

% V_{DS_Q} (calculated) = _____

h. Again borrow the same 2N4416 JFET from the group next to you and insert in your network using the resistors chosen for your design. Measure the resulting levels of V_{DS_Q} and V_{R_D} and calculate I_{D_Q} using the measured resistor value. Record the results below.

V_{DS_Q} (measured) = _____

I_{D_Q} (calculated) = _____

In addition, record the levels of I_{DSS} and V_P for the borrowed JFET.

I_{DSS} (borrowed JFET) = _____

V_P (borrowed JFET) = _____

How do the values of V_{DS_Q} and I_{D_Q} compare with the specified design levels using the borrowed JFET? Use the recorded values of I_{DSS} and V_P for the borrowed JFET to help explain any variations.

How do the overall results for this configuration compare to those obtained for the fixed-bias configuration when the JFETs are interchanged? Does one configuration appear to be less sensitive to changes in JFETs of the same nameplate nomenclature?

Part 4. Problems and Exercises

1. Using the average values of $I_{DSS} = 10$ mA and $V_P = -3.5$ V from the specification sheets for the 2N4416 JFET, redesign the self-bias configuration of Fig. 14.2 for the specified Q-point.

R_D (commercial value) = _____
R_S (commercial value) = _____

Are the values of R_D and R_S within 20% of the design values of Part 2? Comment accordingly.

2. Using the average values of $I_{DSS} = 10$ mA and $V_P = -3.5$ V from the specification sheets for the 2N4416 JFET, redesign the voltage-divider configuration of Fig. 14.4 for the specified Q-point.

R_D (commercial value) = _____
R_S (commercial value) = _____
R_1 (commercial value) = _____
R_2 (commercial value) = _____

Are the values for R_D, R_S, R_1, and R_2 within 20% of the design values of Part 3? Comment accordingly.

3. How else may a designer cope with the variation in I_{DSS} and V_P for JFETs with the same nameplate nomenclature?

Part 5. Computer Exercises

PSpice Simulation 14-1

Design the self-bias circuit shown having the stated criteria:

$$I_{DQ} = 6 \text{ mA}$$
$$V_{DSQ} = 10 \text{ V}$$

PSpice Simulation 14-1

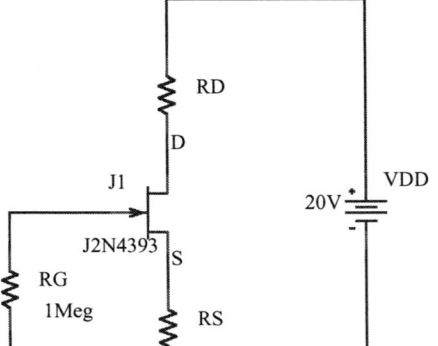

1. For this JFET, use the values of V_P and I_{DSS} as previously obtained from the transfer characteristic in Experiment 12.

2. A 10% positive or negative deviation of V_{DSQ} is allowable.

3. The maximum power to the JFET is not to exceed 100 mW.

4. Run a bias point analysis to check the design.

5. If the criteria are not met, repeat the design procedure and recheck it.

6. Repeat the process until a satisfactory design results.

PSpice Simulation 14-2

The voltage-divider circuit shown is designed to operate with either the J2N4393 or the J2N3819 JFET. It is required that the absolute value of the percent deviation of V_{DS} and I_D be no more than 10%.

1. Run a bias point analysis with the J2N4393 in place.

2. Repeat the bias point analysis with the J2N3819 in place.

3. For both runs, obtain the circuit voltages and the drain currents.

4. From the data, calculate the percent deviations for V_{DS} and I_D for the two circuits. The data from the J2N4393 is used as the standard.

5. Are the values of the calculated deviations within the required 10% limit?

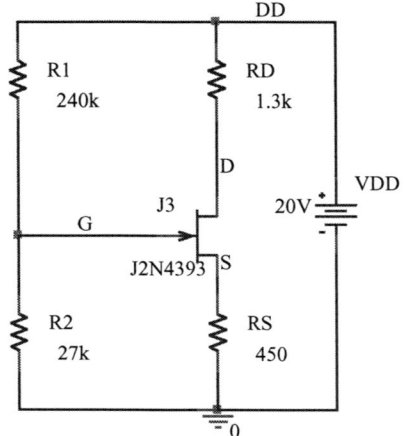

PSpice Simulation 14-2: Voltage divider bias

Name _____
Date _____
Instructor _____

EXPERIMENT 15

Compound Configurations

OBJECTIVES

1. To measure the bias voltages and currents of multistage systems.
2. To demonstrate the independence of the DC voltages and currents of one stage in a capacitively coupled system with the DC voltages and currents of any other stage of the system.
3. To measure the bias voltages and currents of a multistage DC coupled system.

EQUIPMENT REQUIRED

Instrument

DMM

Supplies

DC power supply
9 V battery with snap-on leads

Components

Resistors

(1) 470-Ω
(2) 1-kΩ
(1) 1.2-kΩ
(1) 2.4-kΩ
(1) 2.7-kΩ
(1) 4.7-kΩ

(2) 15-kΩ

(1) 1-kΩ potentiometer

Capacitor

(1) 0.1-μF

Transistors

(2) 2N3904 BJT

(1) 2N4416 JFET

EQUIPMENT ISSUED

Item	Laboratory serial no.
DMM	
DC power supply	

RÉSUMÉ OF THEORY

Typical electronic amplifying systems consist of several transistor stages connected together. The amplification purpose dictates the nature of the interconnection between the stages. If an amplifier is required to amplify a signal containing frequencies well above 0 Hz, the method of coupling most commonly employed is AC coupling. It consists of connecting a capacitor between the collector of one stage and the base of the next stage. In this fashion, the AC component of the collector output voltage is connected into the base of the next stage, while the DC component of the collector voltage is blocked from reaching that base due to the capacitor. In effect, relative to any DC voltages and currents, stages so coupled are isolated from each other. This makes the DC analysis of even the most complex systems relatively easy since each stage can be analyzed independently. In this experiment, the DC biasing levels of various stages of an amplifier are measured. It is then demonstrated that the capacitively coupled stages do not affect each other's DC voltages and currents. Both an exchange of position of the stages in the system, and a change in the biasing network of a transistor are used in that demonstration.

The second coupling system investigated during this experiment is a DC-coupled system. Such systems are used when very-low-frequency components of a signal and even its DC component need to be amplified. A direct connection is made between the collector of a stage and the base of the next. Here it will be demonstrated that any change in the DC voltages and currents in one stage affects the DC voltages and currents in another stage. A technique of changing the bias network on one stage will be used to show the DC dependence of the two stages used in this experiment.

The third compound bias circuit will include a BJT-JFET combination to demonstrate analysis techniques employed for such configurations. The coupling will be direct-coupled to permit a full investigation of the interaction between active devices.

PROCEDURE

Part 1. Determining the BJT (β) and JFET (I_{DSS} and V_P) Parameters

This part of the experiment will determine the BJT and JFET parameters to be employed in the analysis of each compound configuration.

a. To determine the β for each BJT transistor, construct the network of Fig. 15.1 and record the measured resistor values.

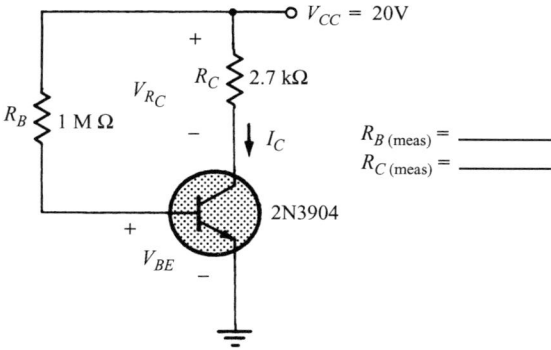

$R_{B \text{ (meas)}}$ = _____

$R_{C \text{ (meas)}}$ = _____

Figure 15-1 Determining β.

Energize the network and measure the voltages V_{BE} and V_{R_C}. Using the measured values of R_B and R_C, calculate the levels of I_B and I_C using the equations $I_B = (V_{CC} - V_{BE})/R_B$ and $I_C = V_{R_C}/R_C$. Then calculate β from $\beta = I_C/I_B$ and record below.

β_1 (calculated) = _____

Replace the 2N3904 with the other 2N3904 BJT transistor and determine its level of β. Insert its level below. Throughout the experiment be sure to identify which transistor has which level of β.

β_2 (calculated) = _____

b. To determine I_{DSS} and V_P for the JFET, construct the network of Fig. 15.2 and record the measured value of R_D.

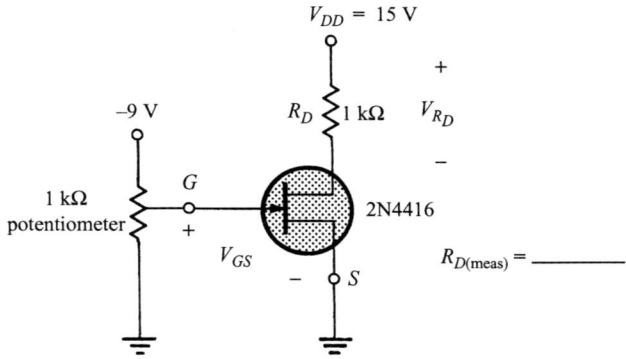

Figure 15-2 Determining I_{DSS} and V_P.

Set V_{GS} to 0 V and measure V_{R_D}. Calculate $I_D = I_{DSS} = V_{R_D}/R_D$ using the measured resistor value and record below.

I_{DSS} (calculated) = _____

Make V_{GS} increasingly negative until V_{R_D} = 1 mV (and $I_D = V_{R_D}/R_D \cong 1$ μA). Since I_D is small ($I_D \cong 0$ A), the resulting value of V_{GS} is the pinch-off voltage V_P. Record below.

V_P (measured) = _____

Part 2. Capacitive-Coupled Multistage System with Voltage-Divider Bias

In this part, bias voltages and currents of a capacitively-coupled two-stage amplifier system are measured. The DC isolation between the two stages will be demonstrated.

    a. Construct the circuit of Fig. 15.3 using a 2N3904 transistor for each stage.

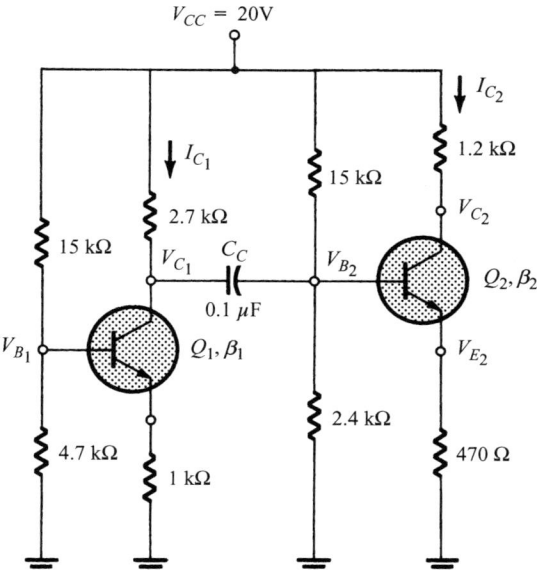

Figure 15-3 AC-coupled multistage amplifier.

b. Using the β values determined in Part 1, calculate the voltage levels V_{B_1}, V_{C_1}, V_{B_2}, and V_{C_2} using measured resistor values. The coupling capacitor C_C will assume the "open-circuit" state for DC conditions. Record the results in Table 15.1.

TABLE 15.1

	V_{B_1}	V_{C_1}	V_{B_2}	V_{C_2}
Calculated Values				
Measured Values				
% Difference				

c. Energize the network of Fig. 15.3 and measure the voltages V_{B_1}, V_{C_1}, V_{B_2}, and V_{C_2} and insert in Table 15.1.

d. Calculate the percent differences between the calculated and measured values using the following equation and record in Table 15.1.

$$\% \text{ Difference} = \frac{|V_{(\text{calc})} - V_{(\text{meas})}|}{|V_{(\text{calc})}|} \times 100\% \qquad (15.1)$$

e. Even though commercial resistor values were employed, are the percent differences in general less than 10%? If not, can you comment on why the difference was so large?

f. Compare the measured values of V_{C_1} and V_{B_2}. Do they confirm the fact that the capacitor C_C assumes an open-circuit state for DC conditions? In other words, for DC conditions, are the two voltage-divider configurations isolated?

Part 3. DC-Coupled Multistage Systems

In this part the bias voltages of a DC-coupled two-stage transistor amplifier will be calculated, measured, and compared. The primary purpose is to demonstrate that the DC levels of one stage will have a direct effect on the DC levels of the other stage.

Exp. 15 / Procedure

a. Construct the network of Fig. 15.4 using the 2N3904 transistors.

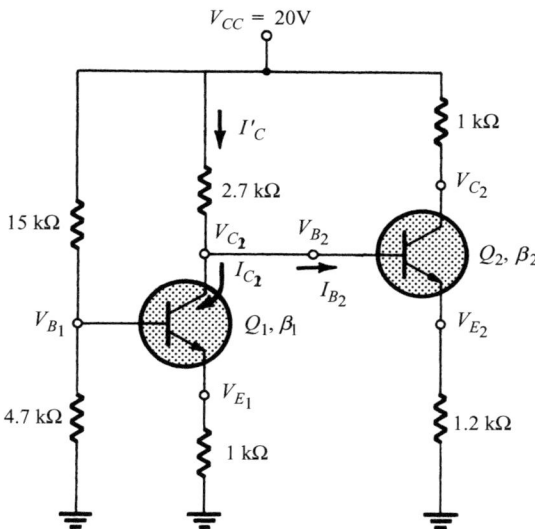

Figure 15-4 DC-coupled multistage amplifier.

b. Using the β values determined in Part 1, calculate the voltage levels V_{B_1}, V_{C_1}, V_{B_2}, and V_{C_2} using measured resistor values. In this case, proceed by first finding V_{B_1}, then V_{E_1}, I_{E_1}, and I_{C_1} followed by V_{C_1} assuming $I'_C \cong I_{C_1} \gg I_{B_2}$. Once $V_{C_1} = V_{B_2}$ is known, V_{E_2} and the remaining unknowns can be found. Insert the results in Table 15.2.

TABLE 15.2

	V_{B_1}	V_{C_1}	V_{B_2}	V_{C_2}
Calculated Values				
Measured Values				
% Difference				

    c. Energize the network of Fig. 15.4 and measure the voltages V_{B_1}, V_{C_1}, V_{B_2}, and V_{C_2} and insert in Table 15.2.

    d. Calculate the percent differences between the calculated and measured values using Eq. 15.1 and record the calculated results in Table 15.2.

    e. Even though commercial resistor values were employed, are the percent differences in general less than 10%? If not, can you comment on why the difference was so large?

    f. Compare the measured values of V_{C_1} and V_{B_2}. Are they equal as expected? Comment on how the DC voltage levels of one stage directly affected the DC voltage levels of the other stage.

Part 4. A BJT-JFET Compound Configuration

A compound configuration with both BJT and JFET transistors will now be examined from a DC viewpoint. The configuration of Fig. 15.5 is direct-coupled, resulting in a direct link in DC levels between the two transistors.

    a. Construct the network of Fig. 15.5 using a 2N3904 BJT transistor and a 2N4416 JFET transistor.

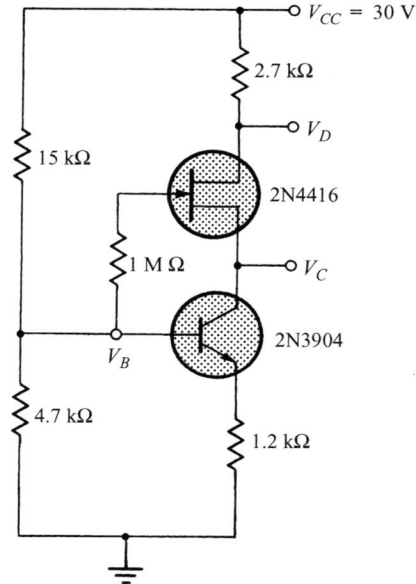

Figure 15-5 BJT-JFET compound configuration.

    b. Using the β, I_{DSS}, and V_P levels determined in Part 1, calculate the DC levels of V_B, V_C, and V_D using measured resistor values. In this case initiate your analysis by first finding V_B and then the level of I_C. Then determine the level of V_D. Using Shockley's equation, calculate the level of V_{GS} followed by V_C. Record the results in Table 15.3.

TABLE 15.3

	V_B	V_D	V_C
Calculated Values			
Measured Values			
% Difference			

c. Energize the network of Fig. 15.5 and measure the voltages V_B, V_D, and V_C and insert in Table 15.3.

d. Calculate the percent differences between the calculated and measured values using Eq. 15.1 and record the calculated results in Table 15.3.

e. Even though commercial resistor values were employed, are the percent differences in general less than 10%? If not, can you comment on why the difference was so large?

f. Determine the voltage V_{GS} from the measurements of Table 15.3. How does it compare to your calculated value from Part 4(**b**)?

V_{GS} (measured) = _____
V_{GS} (calculated) = _____

g. Determine the voltage V_{R_D} from measured values and calculate the drain current from Ohm's law using the commercial resistor value. How does the measured value of I_D compare to the calculated value of Part 4(**b**)?

I_D (measured) = _____
I_D (calculated) = _____

h. Using $V_{BE} = 0.7$ V, calculate the voltage V_E from the measured values of Table 15.3 and then calculate I_C using the commercial resistor value. How does the measured value of I_C compare to the measured value of I_D in Part 4(**g**)?

I_C (measured) = _____

Part 5. Problems and Exercises

1. **a.** For the network of Fig. 15.3 how will the level of V_B and V_C for each transistor change if the two voltage-divider configurations are interchanged?

   b. Will the level of V_B and V_C for each transistor of Fig. 15.3 change if the resistive components maintain their current positions and the transistors are interchanged? Why?

2. Will there be a significant change in the level of V_{E_2} for the network of Fig. 15.4 if the resistive components maintain their current positions and the transistors are interchanged? Support your conclusions with numerical calculations.

3. Remove the 1-MΩ resistor and interchange the positions of the BJT and JFET of Fig. 15.5. Calculate the resulting level of V_B, V_D, and V_C and compare to the levels of Part 4(**b**). Have they changed considerably? Was the change in level expected? Why?

Part 6. Computer Exercises

PSpice Simulation 15-1

The PSpice circuit shown is that of Figure 15-3. Run a bias point analysis and obtain all the voltages and the currents for the two stages. Suggestion: Construct this circuit and run the analysis before the laboratory exercise is performed. The data from the PSpice analysis can be used as a reference against which laboratory data is compared.

PSpice Simulation 15-1: AC coupled multistage amplifier

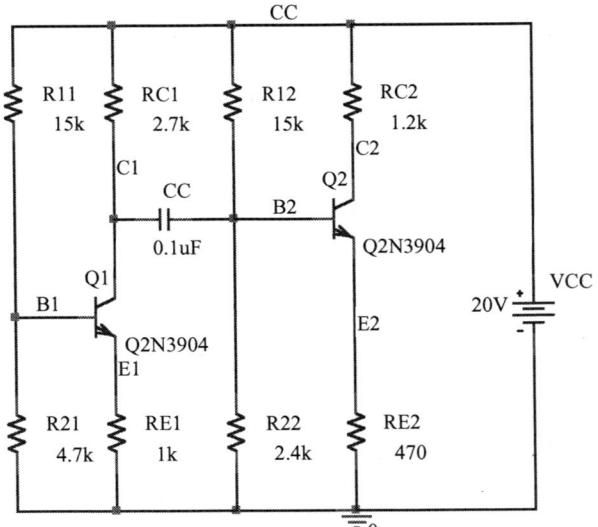

At the completion of the analysis, reverse the stages as shown in the circuit below. Perform a bias point analysis and obtain all the circuit voltages and currents.

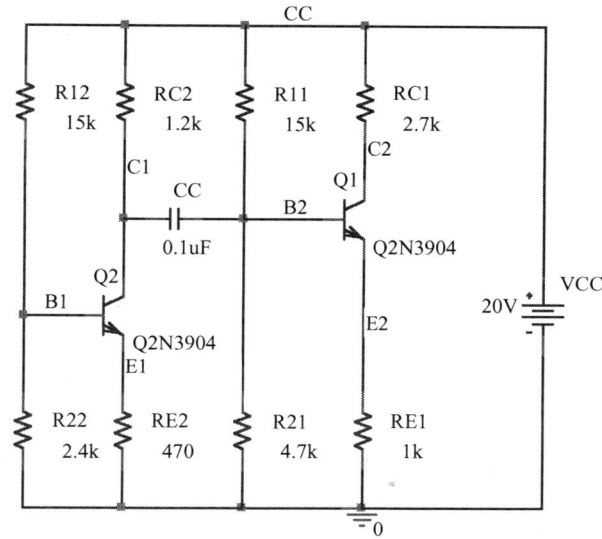

PSpice Simulation 15-1: AC coupled multistage amplifier
Stages interchanged

Answer the following questions:

1. In the first circuit, what is the collector voltage of stages 1 and 2?

2. In the second circuit, what is the collector voltage of stages 1 and 2?

3. Were the collector voltages interchanged when compared to the first circuit?

4. In the first circuit, what are the base and emitter voltages of stage 1?

5. In the second circuit, what are the base and emitter voltages of stage 2?

6. Were these voltages interchanged when compared to the first circuit?

7. In the first circuit, what is the collector current of stages 1 and 2?

8. In the second circuit, what is the collector current of stages 1 and 2?

9. Were the collector currents interchanged when compared to the first circuit?

10. In general, was the DC biasing of either of these stages dependent upon that of the other stage?

11. What accounts for your answer in question 10?

PSpice Simulation 15-2

The PSpice circuit shown is that of Figure 15-4. Run a bias point analysis and obtain all the voltages and the currents for the two stages. Suggestion: Construct this circuit and run the analysis before the laboratory exercise is performed. The data from the PSpice analysis can be used as a reference against which laboratory data is compared.

PSpice Simulation 15-2: DC coupled multistage amplifier

At the completion of the analysis, reverse the stages as shown in the circuit below. Perform a bias point analysis and obtain all the circuit voltages and currents.

PSpice Simulation 15-2: DC coupled multistage amplifier
Stages interchanged

Answer the following questions:
1. In the first circuit, what is the collector voltage of stages 1 and 2?

2. In the second circuit, what is the collector voltage of stages 1 and 2?

3. Were the collector voltages interchanged when compared to the first circuit?

4. In the first circuit, what are the base and emitter voltages of stage 1?

5. In the second circuit, what are the base and emitter voltages of stage 2?

6. Were these voltages interchanged when compared to the first circuit?

7. In the first circuit, what is the collector current of stages 1 and 2?

8. In the second circuit, what is the collector current of stages 1 and 2?

9. Were the collector currents interchanged when compared to the first circuit?

10. In general, was the DC biasing of either of these stages dependent upon that of the other stage?

11. What accounts for your answer in question 10?

Name _____
Date _____
Instructor _____

EXPERIMENT 16

Measurement Techniques

OBJECTIVES

1. To measure the AC and DC amplitudes of a voltage waveform with an oscilloscope.
2. To measure the AC and DC amplitudes of a voltage waveform with a digital multimeter.
3. To measure the period and frequency of periodic voltage waves with an oscilloscope.
4. To measure the frequency of periodic voltage waves with a frequency counter.
5. To measure the phase shift between two sinusoidal voltage waves with an oscilloscope.
6. To study the effect of instrument loading on the voltage measurements in a circuit.

EQUIPMENT REQUIRED

Instruments

Dual-trace oscilloscope
  (Single-trace will limit investigation to Parts 1, 2, and 4)
DMM
Frequency counter

Supplies

DC power supply
Signal generator

Components

Resistors

(2) 1-kΩ

(1) 2-kΩ
(1) 3.9-kΩ
(2) 1-MΩ

Capacitor

(1) 0.1-μF

EQUIPMENT ISSUED

Item	Laboratory serial no.
Oscilloscope	
DMM	
DC power supply	
Signal generator	
Frequency counter	

RÉSUMÉ OF THEORY

This experiment will be an introduction to the measuring instrumentation commonly used to measure DC and AC quantities. Specifically, the oscilloscope and the digital multimeter will be used to measure both the AC and DC components of a voltage waveform. The oscilloscope is basically a voltage-measuring device. It measures the amplitude of any periodic voltage in terms of its peak-to-peak values. By contrast, the digital multimeter measures the rms value of a periodic wave. Note, however, that some digital multimeters measure the rms value of a sinusoidal wave only.

The oscilloscope can also be used to measure the period, and consequently the frequency, of a periodic wave. In the case of a sine wave, it will measure the period of that wave; if a pulse is applied, it will allow for the determination of its fundamental frequency from its period. The frequencies determined from the oscilloscope measurements will be compared with those made with a frequency counter.

It is important in the use of a particular measuring instrument to note any possible effect on the measurements taken. To demonstrate this, a circuit is used which has a low impedance compared to the input impedance of the oscilloscope. Thus the circuit voltage measured approaches its theoretical value. However, when the circuit impedance is increased so that it more nearly approaches that of the oscilloscope, serious measurement errors are introduced. To overcome these errors, 10:1 test probes are used.

PROCEDURE

Part 1. AC and DC Voltage Amplitude Measurements

a. Construct the circuit of Fig. 16.1. Record the measured resistor values.

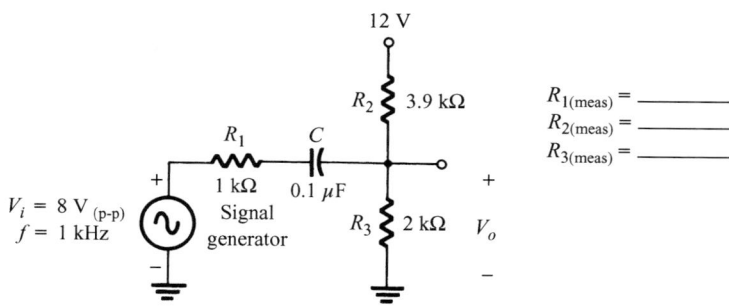

Figure 16-1 AC and DC voltage measurements.

b. Connect the oscilloscope to measure the voltage V_i. For the channel being used, set the AC-GND-DC switch to the GND position and set the horizontal line in the middle of the screen. Then return the AC-GND-DC switch to the AC position.

c. Set the vertical sensitivity to 1 V/cm and adjust the amplitude control of the signal generator until $V_i = 8$ V_{p-p} at a frequency of 1 kHz. Use a horizontal sensitivity of 0.2 ms/cm.

d. Set the DC supply to 12 V using the DMM.

The network is now established with both an AC and a DC supply.

DC Measurements

Both the oscilloscope and DMM will now be used to measure the DC levels of Fig. 16.1.

e. Calculate the expected DC voltage level at V_o using the measured resistor values.

V_o (calculated) = _____

f. Use the DMM to measure the DC level of V_o.

V_o (measured) = _____

Determine the percent difference between the calculated and measured values using the following equation:

$$\% \text{ Difference} = \frac{|V_{o(\text{calc})} - V_{o(\text{meas})}|}{|V_{o(\text{calc})}|} \times 100\% \qquad (16.1)$$

% Difference (calculated) = _____

g. Connect the scope to V_o and set the AC-GND-DC switch to the DC position. Using a sensitivity of 1 V/div., determine the shift (in volts) in the positive peak value (referenced to 0 V) from the value established in Part 1(**c**).

shift in V_o (measured) = _____

Was the shift up or down from the center of the screen? What does the shift tell us about the polarity of V_o?

How does the measured shift with the oscilloscope compare with the DC voltage measured with the DMM?

Is the scope or the DMM more accurate for this type of reading? Why?

AC Measurements

Both the oscilloscope and the DMM will now be used to measure the AC levels of Fig. 16.1.

h. Calculate the rms value of the applied voltage V_i.

$V_{i(rms)}$ (calculated) = _____

i. Calculate the expected rms voltage V_o for the network of Fig. 16.1 at a frequency of 1 kHz, using measured resistor values. Be aware that the reactance of the capacitor must be determined and the vector relationship between resistive and reactive elements employed in the determination. For the AC analysis the 12 V supply can be set to zero volts (superposition applies to the DC/AC analysis of the network), resulting in a parallel arrangement for R_2 and R_3.

$V_{o(\text{rms})}$ (calculated) = _____

j. Use the DMM to measure the rms value of V_o.

$V_{o\ (\text{rms})}$ (measured) = _____

Determine the percent difference between the calculated and measured values using Eq. 16.1.

% Diff. (calculated) = _____

k. Connect the oscilloscope to measure V_o and set the AC-GND-DC switch to the AC position. Using an appropriate vertical and horizontal sensitivity, determine the peak-to-peak value of V_o.

$V_{o(\text{p-p})}$ (measured) = _____

Calculate the rms value of V_o.

$V_{o(\text{rms})}$ (measured) = _____

Determine the percent difference between the calculated and measured values using Eq. 16.1.

% Difference (calculated) = _____

l. Are you satisfied that both the oscilloscope and the DMM can effectively measure the rms values of sinusoidal waveforms? Why?

Part 2. Measurements of the Periods and Fundamental Frequencies of Periodic Voltage Waveforms

In this part of the experiment the oscilloscope will be used to measure the period and to calculate the frequency of a sinusoidal voltage waveform.

    a. Connect the signal generator directly to a vertical channel of the oscilloscope. Set the frequency dial between 1 and 2 kHz *without* taking the time to carefully read the scale and determine which frequency was chosen. Adjust the amplitude control until an 8 $V_{p\text{-}p}$ signal is obtained on the screen.

    An 8 $V_{p\text{-}p}$ sinusoidal signal of unknown frequency is now displayed on the screen. The following is the general procedure to determine the period and frequency of a waveform.

    b. Adjust the horizontal sensitivity until one or two complete cycles of the waveform are displayed on the screen. Record the chosen horizontal sensitivity below.

Horizontal sensitivity = _____

    c. Measure the number of divisions (including fractional parts) encompassed by one full cycle of the waveform on the screen.

Number of divisions = _____

    d. Calculate the period of the waveform by multiplying the horizontal sensitivity by the number of divisions.

Period (T) = _____

    e. The frequency can then be determined using the relationship $f = 1/T$. Calculate the frequency.

Frequency (f) = _____

f. Compare the calculated frequency to the frequency set on the signal generator. Record the set frequency below.

f (dial setting) = _____

g. If the calculated frequency and frequency of the signal generator do not match, can you offer a reason for the difference?

h. Connect the frequency counter to the output voltage terminals and record the displayed frequency.

f (counter) = _____

i. Is the frequency displayed by the counter closer to the frequency calculated using the scope or determined from the dial of the signal generator? Assuming the counter is our best measurement, does a scope or dial setting usually display a more accurate reading of the frequency?

Part 3. Phase-Shift Measurements

a. Construct the network of Fig. 16.2. Record the measured resistor value.

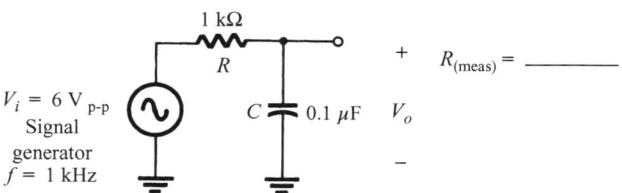

Figure 16-2 Phase-shift measurements.

b. Determine the rms value of the 6 $V_{p\text{-}p}$ signal applied to the input.

$V_{i\ (rms)}$ (calculated) = _____

c. Assuming $V_i = V_i\angle 0°$, calculate $V_o\angle\theta$ at a frequency of 1 kHz.

$V_{o(rms)}$ (calculated) = _____
$V_{o(p\text{-}p)}$ (calculated) = _____
θ = _____

The angle θ is the phase angle between V_i and V_o.

d. Connnect V_i to channel 1 of the oscilloscope and establish V_i as a 6 $V_{p\text{-}p}$, 1 kHz sinusoidal signal balanced (using the AC-GND-DC switch) above and below the center line on the screen using a vertical sensitivity of 1 V/cm. Adjust the waveform so the intersection of the positive slope with the center line occurs at the intersection of one of the vertical grid lines as shown in Fig. 16.3.

e. Connect channel 2 to V_o and, using the same vertical sensitivity of 1 V/cm, superimpose V_o on V_i. Be sure both V_i and V_o are balanced above and below the center line using the GND position of the AC-GND-DC switch for each channel.

f. Count the number of horizontal divisions between positive slopes of V_o and V_i as shown in Fig. 16.3 and label the result A. The separation A represents the phase shift between V_o and V_i.

A (number of divisions) = _____

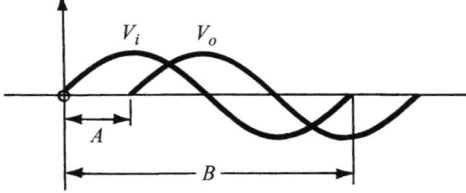

Figure 16-3 Determining the phase shift.

g. Count the number of divisions encompassed by one full cycle of the waveforms and label the result B (note Fig. 16.3).

B (number of divisions) = _____

Exp. 16 / Procedure

h. The phase angle in degrees is determined by using the following equation:

$$\theta = \frac{A}{B} \times 360° \qquad (16.2)$$

Using Eq. 16.2, calculate the phase angle between V_o and V_i for the network of Fig. 16.2.

θ (calculated from time delay) = _____

How does the phase angle measured in Part 3(**h**) compare to the phase angle calculated in Part 3(**c**)?

i. How does the peak-to-peak value of V_o compare to the calculated value of Part 3(**c**)?

j. If V_o crosses the axis with a positive slope to the right of V_i, V_o lags V_i by the angle θ. For the network of Fig. 16.2 does V_o lead or lag V_i? Is the result expected? Why?

k. The phase relationship between V_i and V_R can be obtained by interchanging the positions of the capacitor and resistor. The change in location is required to ensure a common ground between waveforms viewed on the scope.

Interchange the positions of the resistor and capacitor of Fig. 16.2 and calculate the magnitude and angle of V_R assuming $V_i = V_i\angle 0°$.

$V_{R(\text{rms})}$ (calculated) = _____
$V_{R(\text{p-p})}$ (calculated) = _____
θ = _____

1. Use the oscilloscope to measure the magnitude of V_R and V_i. Also indicate if V_o leads or lags V_i.

$V_{R\ (\text{p-p})}$ (measured) = _____
$V_{i\ (\text{p-p})}$ (measured) = _____
θ = _____
lead or lag? _____

How do the measured and calculated results compare?

Part 4. Loading Effects

a. Construct the network of Fig. 16.4. Record the measured values of R_1 and R_2.

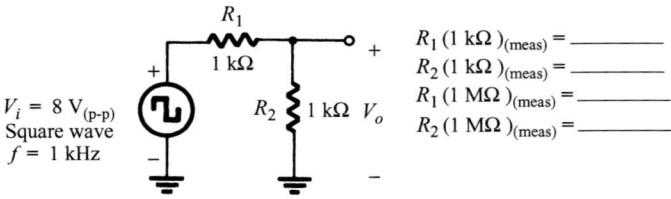

Figure 16-4 Loading effects.

Exp. 16 / Procedure

b. Set V_i to an 8 $V_{p\text{-}p}$ square wave at a frequency of 1 kHz centered on the horizontal center line of the display. Adjust the horizontal sensitivity to show one or two full cycles of V_i.

c. Using the measured resistor values, calculate the peak-to-peak value of V_o.

$V_{o(p\text{-}p)}$ (calculated) = _____

d. Energize the network of Fig. 16.4 and measure the output voltage V_o using the oscilloscope.

$V_{o\ (p\text{-}p)}$ (measured) = _____

How do the results of Parts 4(c) and 4(d) compare?

e. Now replace the two 1-kΩ resistors with 1-MΩ resistors. Insert the measured values of R_1 and R_2.

f. Using the measured resistor values, calculate the peak-to-peak voltage for V_o using the oscilloscope.

$V_{o\ (p\text{-}p)}$ (calculated) = _____

g. Energize the network and measure V_o.

$V_{o\ (p\text{-}p)}$ (measured) = _____

How do the results of Parts 4(f) and 4(g) compare?

h. It is expected that the results of Part 4(**g**) will reveal that the measured and calculated values of V_o do not compare as they did for Parts 4(**c**) and 4(**d**). The change in response is due to the loading by the scope on the circuit when applied to measure V_o. In Fig. 16.5 an additional resistor has been added to that of Fig. 16.4 to represent the loading by the scope on the circuit.

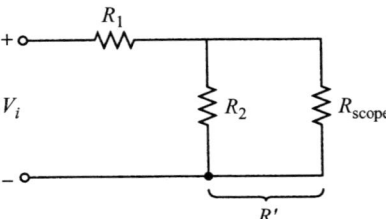

Figure 16-5 Loading by the scope.

Using the measured levels of V_o and V_i, the magnitude of R_{scope} can be obtained by solving for R_{scope} in the following equation obtained using the voltage-divider rule.

$$R' = \frac{R_2 R_{scope}}{R_2 + R_{scope}} = \frac{R_1}{\frac{V_i}{V_o} - 1} \qquad (16.3)$$

Using the measured levels of V_o and V_i, determine R_{scope} using Eq. 16.3 and the measured values of R_1 and R_2.

R_{scope} (calculated) = _____

If the input impedance of the oscilloscope is known, compare it to the calculated value.

i. If R_1 is maintained at 1 MΩ and R_2 replaced by a 1-kΩ resistor, use the results of Part 4(**h**) to calculate the expected level of V_o.

$V_{o(p\text{-}p)}$ (calculated) = _____

j. Energize the network and measure the resulting peak-to-peak value of V_o.

$V_{o(p\text{-}p)}$ (measured) = _____

k. How do the results of Parts 4(**i**) and 4(**j**) compare?

Part 5. Problems and Exercises

1. For the network of Fig. 16.1 is it reasonable to assume the capacitor is simply an "open-circuit" for DC conditions and a "short-circuit" for AC conditions? Do the measured values of the experiment support your conclusions? Why?

2. In general, should a DMM or an oscilloscope be used for DC measurements? Why? When is it advantageous to use an oscilloscope?

3. a. What are the relative advantages of using a DMM over an oscilloscope for measuring AC quantities?

b. What are some of the relative advantages of using an oscilloscope over a DMM for measuring AC quantities?

4. A sinusoidal signal occupies 5 horizontal divisions with the horizontal sensitivity set at 0.1 ms/div. What are the period and frequency of the waveform?

T (calculated) = _____

f (calculated) = _____

5. Determine the phase shift between the waveforms of Fig. 16.6. For the phase angle chosen, which leads which?

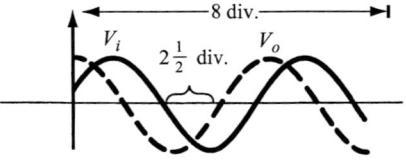

Figure 16-6

θ (calculated) = _____

6. Derive Eq. 16.3.

Part 6. Computer Exercises

PSpice Simulation 16-1

This circuit is the same as that shown in Figure 16-1. We shall run a Time Domain (transient) analysis to obtain the same data as from the experimental measurements. It is recommended that this simulation be done in advance of the laboratory exercise so that the student has a reference against which he or she can compare the experimentally obtained data.

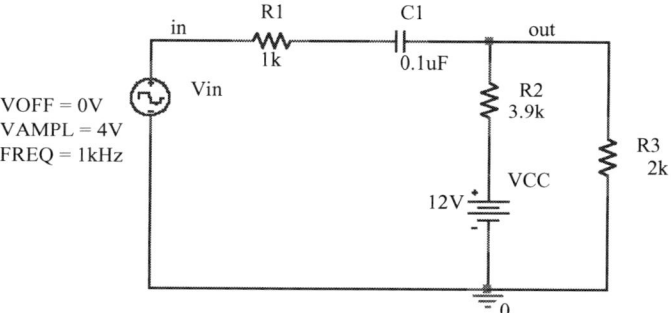

PSpice Simulation 16-1: AC and DC measurements

For this circuit, begin the simulation with a time (transient) analysis of 2 milliseconds duration. This will put two complete cycles of the input and output voltage on the Probe screen. Perform the requested operations and from the data obtained, answer the stated questions:

1. Describe the appearance of the two voltage traces on the Probe screen.

2. Obtain their peak-to-peak voltages using the two Probe cursors.

3. Set the VCC source equal to zero volts and repeat the analysis. The two voltage sources should be centered on the 0-Y axis.

4. Obtain the time delay between the two voltage waveforms. This is most easily done by intercepting two consecutive 0-Y axis crossings of the voltages with the two vertical lines of the cursors. The time delay is labeled A in Figure 16-3. B is the time needed for Vin to complete one cycle. The phase angle in degrees is calculated as in Eq. 16.2.

5. What is the value of the calculated phase shift?

6. Which voltage is leading?

7. We shall next obtain the rms value of the two voltages. To obtain the rms value of the voltages, perform a time (transient) analysis of 20 milliseconds duration. Keep VCC at zero volts.

8. Obtain the Probe traces for Vin and Vout. Use these expressions: for Vin: RMS(V(Vin)); for Vout: RMS(V(out)).

9. Calculate the rms value of these voltages from their peak values.

10. Are the results of steps 8 and 9 in agreement?

11. To obtain the DC value of Vout, run a time (transient) analysis of 20 milliseconds duration. Restore VCC to 12 volts.

12. Obtain the Probe trace for Vout using this expression: AVG(V(out))

13. Compute the DC value of Vout by using the voltage divider formed by R3 and R2.

14. How does this computed value compare with that obtained from the PSpice simulation?

PSpice Simulation 16-2

In the circuit shown, the voltage across R2 is to be measured with a VOM with a sensitivity of 20,000 Ω/volt and with a DMM having an internal resistance of 10 MΩ. Selecting the 10 V scale on the VOM results in an Rmeter of 200 kΩ in parallel with R2. Selecting the DMM places a 10-MΩ resistance in parallel with R2. Perform a bias point analysis with the VOM and obtain the voltage across R2. Repeat the bias point analysis with the DMM and obtain the voltage across R2.

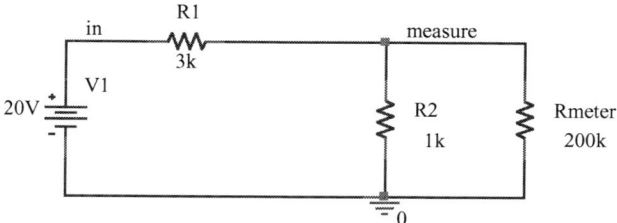

PSpice Simulation 16-2: Loading effects

Replace R1 with a 300-kΩ resistance. Replace R2 with a 100-kΩ resistance. Repeat the bias point analysis as above. Enter the obtained data for the two runs in the chart below:

Meter Used	R2	V(R2) ideal	V(R2) measured	% change
VOM	1 kΩ			
DMM	1 kΩ			
VOM	100 kΩ			
DMM	100 kΩ			

Use the ideal value of V(R2) as the standard of comparison.

From the above compilation, answer the following questions:

1. Which meter, in what circuit, showed the largest loading effect?

2. Which meter, in what circuit, showed the least loading effect?

3. If a loading effect, resulting in less than 1% reduction of V(R2), is mandated, in which circuit can the VOM be used?

4. For the same loading effect, in which circuit can the DMM be used?

5. As a consequence of your data, would it be advisable to use a VOM in most electronic circuits?

Name _____
Date _____
Instructor _____

Common-Emitter Transistor Amplifiers

OBJECTIVES

1. To measure AC and DC voltages in a common-emitter amplifier.
2. To obtain measured values of voltage amplification (A_v), input impedance (Z_i), and output impedance (Z_o) for loaded and unloaded operation.

EQUIPMENT REQUIRED

Instruments

Oscilloscope
DMM
Function generator
DC power supply

Components

Resistors

(2) 1-kΩ
(2) 3-kΩ
(1) 10-kΩ
(1) 33-kΩ

Capacitors

(2) 15-μF
(1) 100-μF

Transistors

(2) NPN (2N3904, 2N2219, or equivalent general purpose transistor)

EQUIPMENT ISSUED

Item	Laboratory serial no.
DC power supply	
Function generator	
Oscilloscope	
DMM	

RÉSUMÉ OF THEORY

The common-emitter (CE) transistor amplifier configuration is widely used. It provides large voltage gain (typically tens to hundreds) and provides moderate input and output impedance. The AC signal voltage gain is defined as

$$A_v = V_o/V_i$$

where V_o and V_i can both be rms, peak, or peak-to-peak values. The input impedance, Z_i, is that of the amplifier as seen by the input signal. The output impedance, Z_o, is that seen looking from the load into the output of the amplifier.

For the voltage-divider DC bias configuration (see Fig. 17.1), all DC bias voltages can be approximately determined without knowing the exact value of the transistor's beta. The transistor's AC dynamic resistance, r_e, can be calculated using

$$r_e = \frac{26(\text{mV})}{I_{E_Q}(\text{mA})} \tag{17.1}$$

AC Voltage Gain: The AC voltage gain of a CE amplifier (under no-load) can be calculated using

$$A_v = \frac{-R_C}{(R_E + r_e)}$$

If R_E is bypassed by a capacitor, use $R_E = 0$ in the above equation.

Thus:
$$A_v = \frac{-R_C}{r_e} \tag{17.2}$$

AC Input Impedance: The AC input impedance is calculated using

$$Z_i = R_1 \| R_2 \| \beta(R_E + r_e)$$

If R_E is bypassed by a capacitor, use $R_E = 0$ in the above equation.

Thus:
$$Z_i = R_1 \| R_2 \| \beta r_e \tag{17.3}$$

AC Output Impedance: The AC output impedance is

$$Z_o = R_C \tag{17.4}$$

Exp. 17 / Procedure

PROCEDURE

Part 1. Common-Emitter DC Bias

a. Record measured values of each resistor in Fig. 17.1.

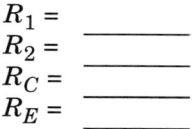

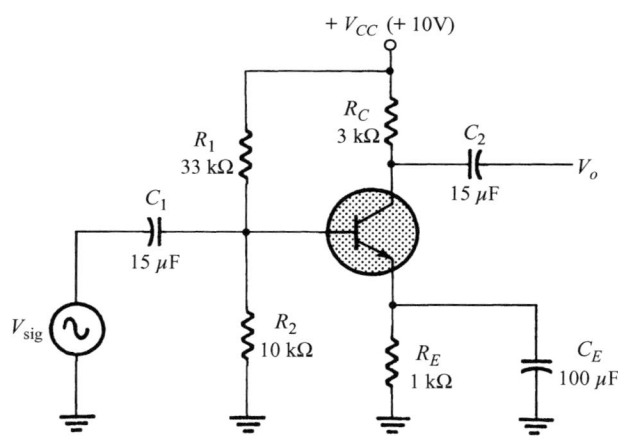

Figure 17-1

b. Calculate DC bias values for the circuit of Fig. 17.1. Record calculated values below.

V_B (calculated) = _____
V_E (calculated) = _____
V_C (calculated) = _____
I_E (calculated) = _____

Calculate r_e using Eq. 17.1 and the calculated level of I_E.

r_e (calculated) = _____

c. Wire up the circuit of Fig. 17.1. Set $V_{CC} = 10$ V. Check the DC bias of the circuit measuring values of

V_B (measured) = _____
V_E (measured) = _____
V_C (measured) = _____

Check that these values compare closely with those calculated in Part 1(**b**). Calculate the DC emitter current using

$$I_E = V_E/R_E$$

$I_E =$ _____

Calculate the AC dynamic resistance, r_e, using the measured value of I_E.

$$r_e = \frac{26(\text{mV})}{I_E(\text{mA})}$$

$r_e =$ _____

Compare r_e with that calculated in Part 1(**b**).

Part 2. Common-Emitter AC Voltage Gain

a. Calculate the amplifier voltage gain for a fully bypassed emitter using Eq. 17.2.

A_v (calculated) = _____

b. Apply an AC input signal, $V_{sig} = 20$ mV, rms at $f = 1$ kHz. Observe the output waveform on the scope to be sure that there is no distortion (if there is, reduce the input signal or check the DC bias). Measure the resulting AC output voltage, V_o, using a scope or a DMM.

V_o (measured) = _____

Calculate the circuit no-load voltage gain using measured values.

$$A_v = \frac{V_o}{V_{sig}}$$

$A_v =$ _____

Compare the measured value of A_v with that calculated in Part 2(**a**).

Part 3. AC Input Impedance, Z_i

a. Calculate Z_i using Eq. 17.3. Use the beta measured with a transistor curve tracer or beta tester, or the nominal listed value in specification sheets (say, $\beta = 150$).

Z_i (calculated) = _____

b. To measure Z_i connect an input measurement resistor, $R_x = 1\ k\Omega$, as shown in Fig. 17.2. Apply input $V_{sig} = 20$ mV, rms. Observe the output waveform with a scope to ensure that no distortion is present (adjust input amplitude if necessary). Measure V_i.

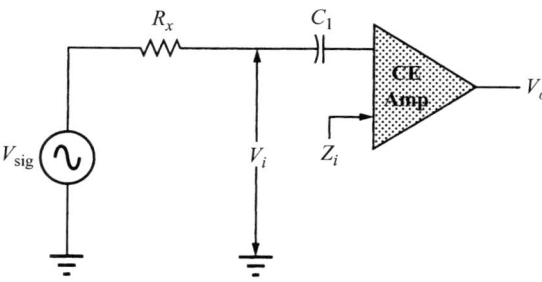

Figure 17-2

V_i (measured) = _____

Solving for V_i using

$$V_i = \frac{V_{sig}}{(Z_i + R_x)} Z_i$$

we get

$$Z_i = \frac{V_i}{(V_{sig} - V_i)} R_x$$

Z_i = _____

Compare the measured value of Z_i with that calculated in Part 3(**a**).

Part 4. Output Impedance, Z_o

a. Calculate Z_o using Eq. 17.4.

Z_o (calculated) = _____

b. Remove the input measurement resistor, R_x. For input of V_{sig} = 20 mV rms, measure the output voltage, V_o. Check the output waveform to ensure that no distortion is present.

V_o [measured] (unloaded) = V_o = _____

Now connect load R_L = 3 kΩ and measure V_o.

V_o [measured] (loaded) = V_L = _____

The output impedance can be obtained from

$$V_L = \frac{R_L}{(Z_o + R_L)} V_o$$

for which

$$Z_o = \frac{V_o - V_L}{V_L} R_L$$

Z_o = _____

Compare the measured value of Z_o with that calculated in Part 4(**a**).

Part 5. Oscilloscope Measurement

Connect the amplifier of Fig. 17.1. For input of $V_{sig} = 20$ mV, p-p, at a frequency of $f = 1$ kHz, sketch the waveforms for V_{sig} and V_o in Fig. 17.3.

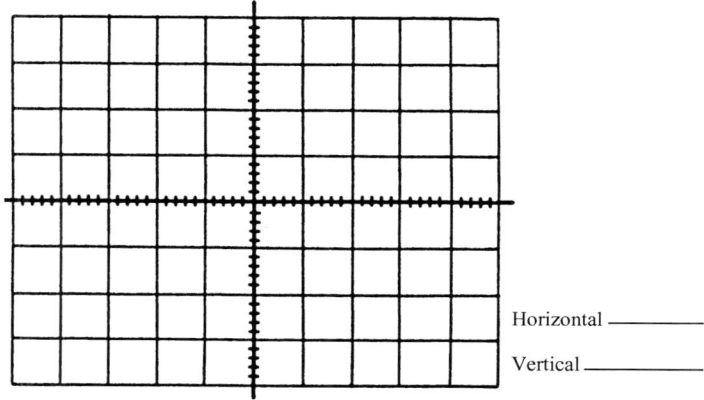

Horizontal _____

Vertical _____

Figure 17-3

Part 6. Computer Exercise

PSpice Simulation 17-1

The common-emitter circuit shown is that of Figure 17-1. The resistor R3 has been added to keep node "out" from floating. Such a condition is not allowed in PSpice. Its addition will not alter the fundamental responses of this circuit. Perform the steps listed on the next page.

PSpice Simulation 17-1: Common Emitter

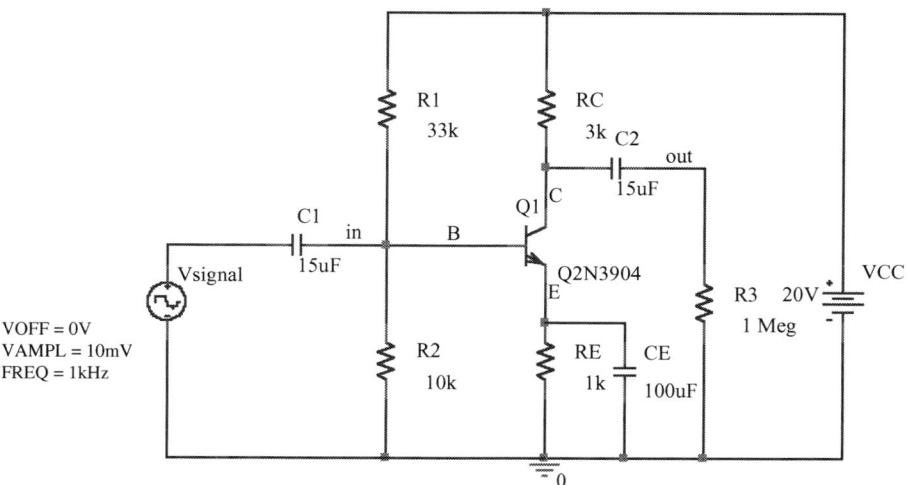

Exp. 17 / Common-Emitter Transistor Amplifiers

1. Perform a bias point analysis to obtain all the DC currents and voltages of this circuit.

2. Calculate the resistance r_e from the obtained data.

3. Perform a Time Domain (transient) analysis for 2 milliseconds and plot the voltages V(Signal) and V(out) on a Probe plot.

4. Measure their peak-to-peak voltages using the Probe cursors.

5. What is the phase shift of the two voltages relative to each other?

6. Can you explain the reason for that phase shift?

7. Calculate the theoretical input impedance of this amplifier.

8. Perform a Time Domain (transient) analysis of 20 milliseconds duration.

9. Plot the ratio of RMS(V(VSignal))/RMS(I(C1)). This ratio is equal to the input impedance of the amplifier.

10. Compare this value with its theoretical one. Are they in agreement?

Modify the above circuit as shown to obtain the output impedance. Note that the Vsignal source has been set to zero volts. The amplitude of the voltage VTest is relatively unimportant, at least in theory!

PSpice Simulation 17-1: Common Emitter

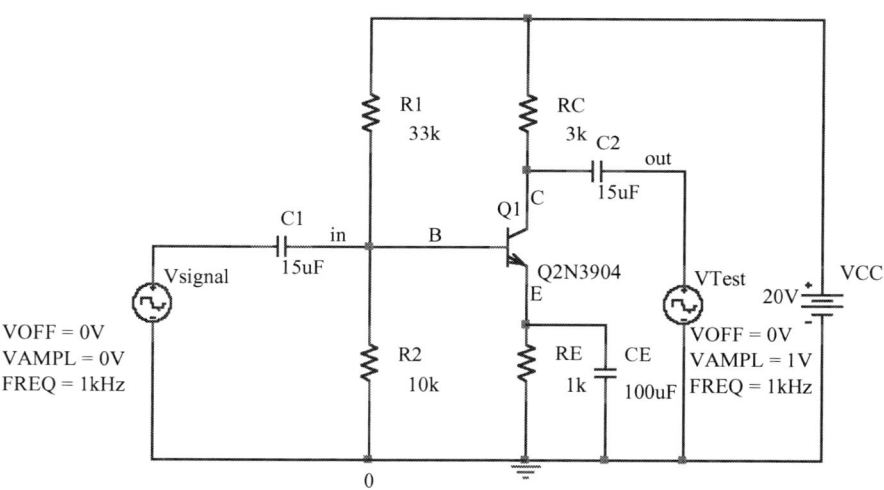

1. Calculate the theoretical value of the output impedance.

2. Perform a Time Domain (transient) analysis of 20 milliseconds duration.

3. Plot the ratio of RMS(V(Test))/RMS(I(C2)). This ratio is equal to the output impedance of the amplifier.

4. Compare this value to its theoretical one. Are they in agreement?

5. If not, what could account for their difference?

EXPERIMENT 18

Common-Base and Emitter-Follower (Common-Collector) Transistor Amplifiers

OBJECTIVES

1. To measure DC and AC voltages in common-base and emitter-follower (common-collector) amplifiers.
2. To obtain measured values of voltage amplification (A_v), input impedance (Z_i) and output impedance (Z_o).

EQUIPMENT REQUIRED

Instruments

Oscilloscope
DMM
Function generator
DC power supply

Components

Resistors

(1) 100-Ω
(1) 1-kΩ
(2) 3-kΩ
(2) 10-kΩ
(1) 33-kΩ
(1) 100-kΩ

Capacitors

(2) 15-μF
(1) 100-μF

Transistors

(2) NPN (2N3904, 2N2219, or equivalent general purpose)

EQUIPMENT ISSUED

Item	Laboratory serial no.
DC power supply	
Function generator	
Oscilloscope	
DMM	

RÉSUMÉ OF THEORY

The common-base (CB) transistor amplifier configuration is used primarily for higher frequency operation. It provides large voltage gain at low input and moderate output impedance. Its voltage gain is

$$A_v = \frac{R_C}{r_e} \tag{18.1}$$

AC Input Impedance: The AC input impedance is

$$Z_i = r_e \quad \text{(grounded based terminal)} \tag{18.2}$$

AC Output Impedance: The AC output impedance is

$$Z_o = R_C \tag{18.3}$$

The common-collector (CC) or emitter-follower (EF) transistor amplifier configuration is used primarily for impedance matching operation. It provides voltage gain near unity and high input and low output impedance.

AC Voltage Gain: The AC voltage gain of a CC amplifier is calculated as

$$A_v = \frac{R_E}{R_E + r_e} \tag{18.4}$$

AC Input Impedance: The AC input impedance is calculated as

$$Z_i = R_1 || R_2 || \beta(R_E + r_e) \tag{18.5}$$

AC Output Impedance: The AC output impedance is

$$Z_o = r_e \tag{18.6}$$

PROCEDURE

Part 1. Common-Base DC Bias

a. Calculate DC bias current and voltages for the circuit of Fig. 18.1. Record calculated values below.

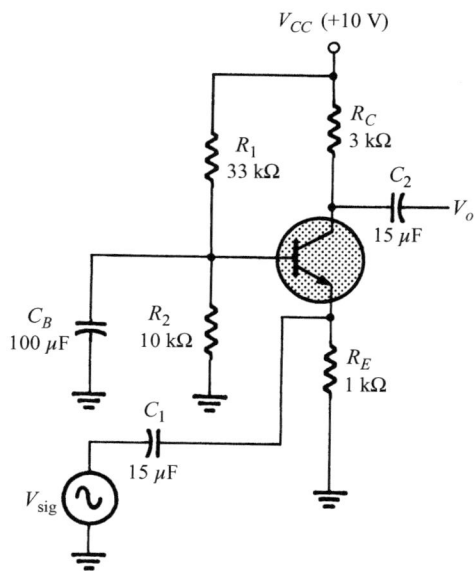

Figure 18-1

V_B (calculated) = _____
V_E (calculated) = _____
V_C (calculated) = _____
I_E (calculated) = _____

Calculate r_e using $r_e = 26(\text{mV})/I_E(\text{mA})$.

r_e (calculated) = _____

b. Wire up the circuit of Fig. 18.1. Set $V_{CC} = 10$ V. Check the DC bias of the circuit measuring values of

V_B (measured) = _____
V_E (measured) = _____
V_C (measured) = _____

Calculate the DC emitter current using

$$I_E = V_E/R_E$$

$I_E =$ _____

Calculate the AC dynamic resistance, r_e.

$$r_e = 26(\text{mV})/I_E \text{ (mA)}$$

$r_e =$ _____

Compare the DC voltages, current I_E, and dynamic resistance r_e calculated in Part 1(a) with the values obtained in Part 1(b).

Part 2. Common-Base AC Voltage Gain

a. Calculate the AC voltage gain of the CB amplifier in Fig. 18.1 using Eq. 18.1.

A_v (calculated) = _____

b. Apply an AC input signal, $V_{sig} = 50$ mV, rms at a frequency of 1 kHz. Measure the resulting AC output voltage, V_o.

V_o (measured) = _____

Calculate the circuit AC voltage gain.

$$A_v = \frac{V_o}{V_{sig}}$$

$$A_v = \underline{\hspace{2cm}}$$

Compare the voltage gain calculated in Part 2(**a**) with that measured in Part 2(**b**).

Using the oscilloscope, observe and sketch the input voltage waveform, V_{sig}, and the output voltage waveform, V_o, in Fig. 18.2.

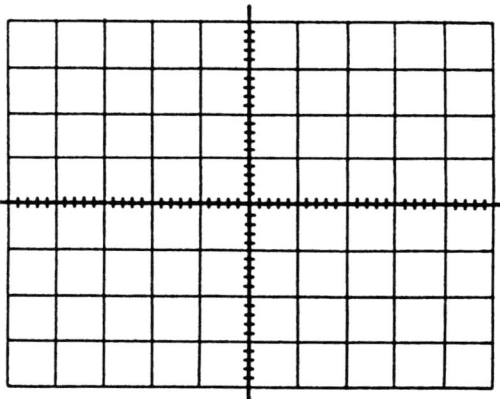

Figure 18-2

Part 3. CB Input Impedance, Z_i

    a. Obtain the AC input impedance of the CB amplifier in Fig. 18.1 using Eq. 18.2.

$$Z_i \text{ (calculated)} = \underline{\hspace{2cm}}$$

b. To measure Z_i connect the input measurement resistor, $R_x = 100\ \Omega$, as shown in Fig. 18.3. Apply input $V_{sig} = 50$ mV, rms at frequency $f = 1$ kHz. Measure V_i.

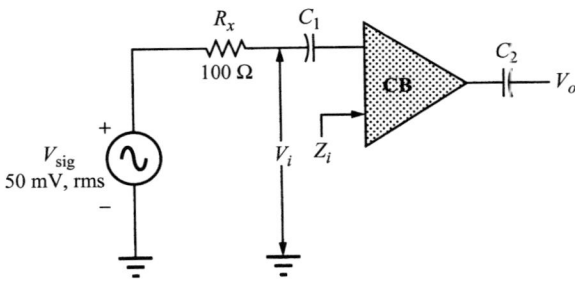

Figure 18-3

V_i (measured) = _____

Calculate using

$$V_i = \frac{Z_i}{(Z_i + R_x)} V_{sig}$$

$$Z_i = \frac{V_i}{(V_{sig} - V_i)} R_x$$

Z_i (measured) = _____

Remove resistor R_x.

Compare the AC input impedance calculated in Part 3(**a**) with that measured in Part 3(**b**).

Part 4. CB Output Impedance, Z_o

a. Determine the AC output impedance of the CB amplifier of Fig. 18.1 using Eq. 18.3.

Z_o (calculated) = _____

b. For an input of $V_{sig} = 20$ mV at a frequency of 1 kHz, rms measure the output voltage, V_o, with no load connected.

V_o (measured) (unloaded) = _____

Now connect load $R_L = 3$ kΩ and measure V_L.

V_L (measured) = _____

The output impedance can be calculated from

$$V_L = \frac{R_L}{(Z_o + R_L)} V_o$$

Hence,

$$Z_o = \frac{V_o - V_L}{V_L} R_L$$

Z_o = _____

Compare the AC output impedance calculated in Part 4(**a**) with the output impedance calculated from measured voltage data in Part 4(**b**).

Part 5. Emitter-Follower DC Bias

a. Calculate DC bias current and voltages for the EF circuit of Fig. 18.4. Record calculated values below.

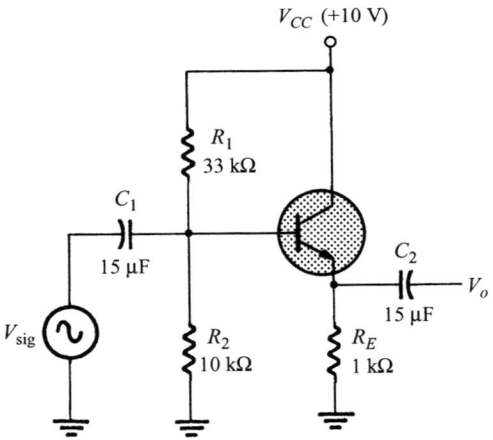

Figure 18-4

V_B (calculated) = _____
V_E (calculated) = _____
V_C (calculated) = _____
I_E (calculated) = _____

Calculate r_e using $r_e = 26(\text{mV})/I_E$ (mA).

r_e(calculated) = _____

b. Wire up the circuit of Fig. 18.4. Set $V_{CC} = 10$ V. Check the DC bias of the circuit measuring values of

V_B (measured) = _____
V_E (measured) = _____
V_C (measured) = _____

Calculate using

$$I_E = \frac{V_E}{R_E}$$

I_E = _____

Calculate the value of r_e using

$$r_e = \frac{26 \text{ mV}}{I_E}$$

$r_e =$ _____

Compare the DC voltages and current calculated in Part 5(**a**) with those measured in Part 5(**b**).

Part 6. Emitter-Follower AC Voltage Gain

a. Calculate the AC voltage gain of an EF amplifier using Eq. 18.4.

A_v (calculated) = _____

b. Apply an AC input signal, $V_{sig} = 1$ V, rms at a frequency of 1 kHz. Measure the resulting AC output voltage, V_o.

V_o (measured) = _____

Calculate the circuit AC voltage gain.

$$A_v = \frac{V_o}{V_{sig}}$$

A_v (measured) = _____

Compare the voltage gain calculated in Part 6(**a**) with that measured in Part 6(**b**).

Observe and sketch the input signal, V_{sig}, and output voltage, V_o,

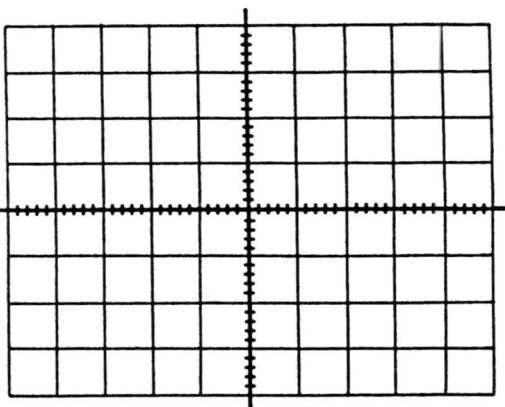

in Fig. 18.5.

Figure 18-5

Part 7. Emitter-Follower (EF) Input Impedance, Z_i

a. Calculate the AC input impedance of an EF amplifier using Eq. 18.5.

Z_i (calculated) = _____

b. To measure Z_i connect the input measurement resistor, $R_x = 10$ kΩ, as shown in Fig. 18.6. Apply input $V_{sig} = 2$ V, rms at frequency $f = 1$ kHz. Measure V_i.

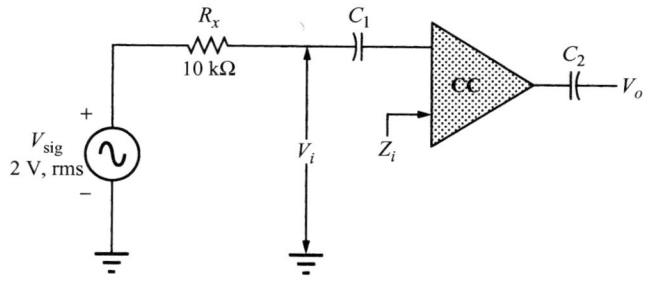

Figure 18-6

V_i (measured) = _____

Calculate Z_i using

$$V_i = \frac{Z_i}{(Z_i + R_x)} V_{sig}$$

$$Z_i = \frac{V_i}{(V_{sig} - V_i)} R_x$$

Z_i = _____

Compare the AC input impedance of a CC amplifier calculated in Part 7(**a**) with that measured in Part 7(**b**).

Part 8. Emitter-Follower (EF) Output Impedance, Z_o

a. Calculate the AC output impedance of a CC amplifier using Eq. 18.6.

Z_o (calculated) = _____

b. For input of $V_{sig} = 20$ mV, rms at frequency $f = 1$ kHz measure the output voltage, V_o.

V_o (measured) = _____

Now connect load $R_L = 100\ \Omega$ and measure V_L.

V_L (measured) = _____

The output impedance is calculated from

$$V_L = \frac{R_L}{(Z_o + R_L)} V_o$$

Hence,

$$Z_o = \frac{V_o - V_L}{V_L} R_L$$

Z_o = _____

Compare the CC output impedance calculated in Part 8(**a**) with that determined in Part 8(**b**).

Part 9. Computer Exercises

PSpice Simulation 18-1

For the common-base amplifier shown, perform a bias point analysis and do the following:

1. Obtain the DC voltages of the amplifier.

2. Obtain the base and collector currents.

3. Compare your PSpice data with the DC voltages and currents obtained experimentally.

Using the equations listed in the "Resumé of Theory" section of Experiment 18:

4. Calculate the value of the dynamic resistance r_e.

5. Calculate the theoretical voltage gain of the amplifier.

6. Calculate the theoretical value of the input impedance.

7. Calculate the theoretical value of the output impedance.

PSpice Simulation 18-1: Common Base Amplifier

Note: Rload has been added to prevent node "out" from floating. Its addition will not significantly affect the operation of this circuit.

Comparison of the Input and the Output Voltages

Next, perform a Time Domain (transient) analysis of 2 milliseconds duration. Vsignal has a peak amplitude of 10 millivolts and a frequency of 1 kHz.

Performing the following steps:

1. Obtain a Probe plot of both the output and the signal voltage. It is best to use two different y-axes for the two voltages because of their different amplitudes.

2. Determine the phase angle between these two voltages.

3. Repeat the Time Domain (transient) analysis for a duration of 20 milliseconds.

4. Obtain the ratio of RMS(V(Vout))/RMS(V(Vsignal: +)). This ratio is equal to the AC gain of this amplifier. Use a cursor to obtain that amplitude.

5. Compare its value to its theoretical value calculated above.

6. Compare its value to that obtained from experimental data.

7. Account for any differences in the AC gains obtained.

Determination of the Input and Output Impedances from PSpice Data

With the Time Domain (transient) analysis still set at 20 milliseconds, obtain the ratio of RMS(V(Vsignal:+))/RMS(I(C1)). This ratio is equal to the input impedance.

1. From its Probe plot, use a cursor to determine its amplitude.

2. Compare this value to its theoretical one.

3. Compare this value to that obtained from laboratory data.

4. Account for any differences in the input impedances obtained.

To obtain the output impedance, modify the amplifier as shown:

PSpice Simulation 18-1: Common Base Amplifier

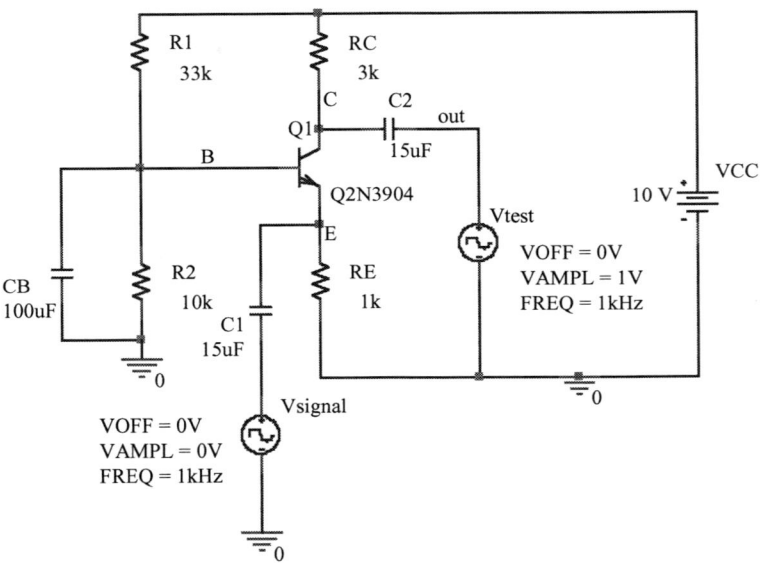

Vsignal is set at 0 volts.
1. Run a Time Domain (transient) analysis of 20 milliseconds duration and obtain the ratio of RMS(V1(Vtest))/RMS(I(C2)). This ratio is equal to the output impedance.

2. Compare this value to its theoretical one.

3. Compare this value to that obtained from laboratory data.

4. Account for any differences in the output impedances obtained.

PSpice Simulation 18-2

For the emitter-follower amplifier shown, perform a bias point analysis and do the following:

1. Obtain the DC voltages of the amplifier.

2. Obtain the base and collector currents.

3. Compare your PSpice data with the DC voltages and currents obtained experimentally.

Using the equations listed in the "Resumé of Theory" section of Experiment 18:

4. Calculate the value of the dynamic resistance r_e.

5. Calculate the theoretical voltage gain of the amplifier.

Exp. 18 / Common-Base and Emitter-Follower (Common-Collector) Transistor Amplifiers

6. Calculate the theoretical value of the input impedance.

7. Calculate the theoretical value of the output impedance.

PSpice Simulation 18-2: Emitter Follower

Comparison of the Input and Output Voltages.

Next, perform a Time Domain (transient) analysis of 2 milliseconds duration. Vsignal has a peak amplitude of 1 volt and a frequency of 1 kHz.

Perform the following steps:
1. Obtain a Probe plot of both the output and the signal voltage. There is no need to use two different y-axes. Why not?

2. What is the phase angle between these two voltages?

3. Repeat the Time Domain (transient) analysis for a duration of 20 milliseconds.

4. Obtain the ratio of RMS(V(Vout))/RMS(V(Vsignal: +)). This ratio is equal to the AC gain of the amplifier. Use a cursor to obtain that amplitude.

5. Compare its value to its theoretical value calculated above.

6. Compare its value to that obtained from experimental data.

7. Account for any differences in the AC gains obtained.

Determination of the Input and Output Impedances Obtained from PSpice Data

With the Time Domain (transient) analysis still set at 20 milliseconds, obtain the ratio of RMS(V(Vsignal:+))/RMS(I(C1)). This ratio is equal to the input impedance.

1. From its Probe plot, use a cursor to determine its amplitude.

2. Compare this value to its theoretical one.

3. Compare this value to that obtained from laboratory data.

4. Account for any differences in the input impedances obtained.

To obtain the output impedance, modify the amplifier as shown:

PSpice Simulation 18-2: Emitter Follower

Vsignal is set at 0 volts.

1. Run a 20 milliseconds Time Domain (transient) analysis of 20 milliseconds duration and obtain the ratio of RMS(V1(Vtest))//RMS(I(C2)). This ratio is equal to the output impedance.

2. Compare this value to its theoretical one.

3. Compare this value to that obtained from laboratory data.

4. Account for any differences in the output impedances obtained.

Name
Date
Instructor

Design of Common-Emitter Amplifiers

OBJECTIVES

1. To design, build, and test a common-emitter amplifier.
2. To calculate and measure DC bias and AC amplification values.

EQUIPMENT REQUIRED

Instruments

Oscilloscope
DMM
Function generator
DC power supply

Components

Resistors

To be selected in design

Capacitors

To be selected in design

Transistors

(1) NPN (2N3904, 2N2219, or equivalent general purpose)

EQUIPMENT ISSUED

Item	Laboratory serial no.
DC power supply	
Function generator	
Oscilloscope	
DMM	

RÉSUMÉ OF THEORY

This lab will design a common-emitter amplifier as shown in Fig. 19.1. The design process begins with a set of specifications that define both the transistor and the circuit operation desired. Fig. 19.1 shows a voltage-divider amplifier with emitter resistor R_E fully bypassed. If possible, a computer should be used to perform the design and test the circuit before it is built. Either PSpice or Microcap II can be used to test any circuit design obtained. Using a 2N3904 (or equivalent transistor), design specifications are:

$\beta = 100$ typical
$I_C(\text{max}) = 200$ mA
$V_{CE}(\text{max}) = 40$ V

The circuit should have the following features:

$V_{CC} = 10$ V
$A_v = 100$ minimum
$Z_i = 1$ kΩ minimum
$Z_o = 10$ kΩ maximum
AC output voltage swing = 3 $V_{p\text{-}p}$ maximum
Load resistance, $R_L = 10$ kΩ minimum

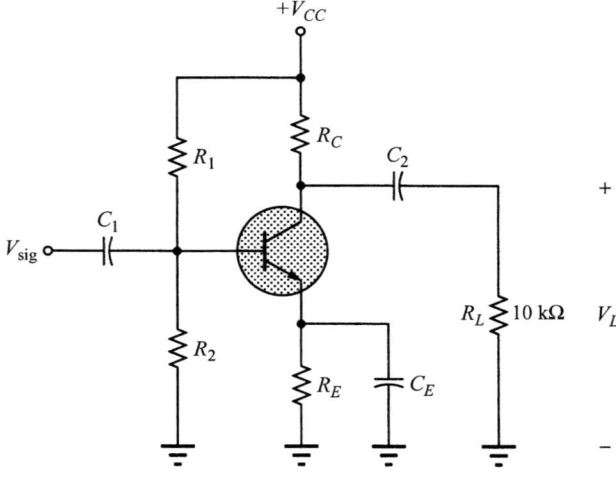

Figure 19-1

PROCEDURE

Part 1. Selection of Components

The CE circuit to be built is that shown in Fig. 19.1. The level of V_{CC} (10 V) is well within the maximum rating of the transistor (V_{CE} = 40 V maximum), and would allow a 3 $V_{p\text{-}p}$ output voltage swing. For the mid-band frequency of f = 1 kHz, capacitor values of $C_1 = C_2$ = 15 µF and C_E = 100 µF would be satisfactory.* For the transistor consider a minimum β = 100 in the design.

a. Select

$$V_E = \frac{V_{CC}}{10} = \frac{10 \text{ V}}{10} = 1 \text{ V}$$

b. Have each laboratory team design for a different value of I_C.** The design shown here is for I_C = 1 mA. For $I_E = I_C$ = 1 mA, the value of R_E should be

$$R_E = \frac{V_E}{I_E} = \frac{1 \text{ V}}{1 \text{ mA}} = 1 \text{ k}\Omega$$

c. Select R_C to bias the circuit at about V_{CE} = 5 V (one-half V_{CC}). Then $V_{RC} = V_{CC} - V_{CE} - V_E$ = 4 V, and

$$R_C = \frac{V_{R_C}}{I_C} = \frac{4 \text{ V}}{1 \text{ mA}} = 4 \text{ k}\Omega \text{ (use 4.1 k}\Omega\text{)}$$

d. Check A_v:

$$r_e = \frac{26 \text{ mV}}{I_E \text{ (mA)}} = \frac{26}{1} = 26 \text{ }\Omega$$

$$|A_v| = \frac{R_C}{r_e} = \frac{4.1 \text{ k}\Omega}{26 \text{ }\Omega} = 158$$

e. Since the input impedance looking into the base of the transistor is βr_e = 100(26 Ω) = 2.6 kΩ, select R_1 and R_2 as large as possible but still sensitive to the DC condition $\beta R_E \geq 10 R_2$ so the system is not loaded down.

Using $\beta R_E \geq 10 R_2$

we find $R_2 \leq \dfrac{\beta R_E}{10} = \dfrac{(100)(1 \text{ k}\Omega)}{10} = 10 \text{ k}\Omega$

$\therefore$ use R_2 = 10 kΩ

---

*For 15 µF: $X_C = 1/(2\pi f C) = 1/[2\pi(1\times10^3)(15\times10^{-6})] = 10.6 \text{ }\Omega$.
For 100 µF: $X_C = 1/(2\pi f C) = 1/[2\pi(1\times10^3)(100\times10^{-6})] = 1.6 \text{ }\Omega$.
**Laboratory group 1 to use I_C = 1 mA, laboratory group 2 to use I_C = 2 mA, etc.

Substituting into the basic equation

$$V_B = \frac{R_2 V_{CC}}{R_1 + R_2} = V_E + 0.7 \text{ V} = 1 \text{ V} + 0.7 \text{ V} = 1.7 \text{ V}$$

$$V_B = \frac{10 \text{ k}\Omega(10 \text{ V})}{R_1 + 10 \text{ k}\Omega} = 1.7 \text{ V}$$

and $100 \text{ k}\Omega = 1.7 R_1 + 17 \text{ k}\Omega$

or $1.7 R_1 = 83 \text{ k}\Omega$

and $R_1 = \dfrac{83 \text{ k}\Omega}{1.7} \cong 48.82 \text{ k}\Omega$ (use 47 kΩ)

f. Check Z_i:

$$Z_i = R_1 \| R_2 \| \beta r_e = 130 \text{ k}\Omega \| 27 \text{ k}\Omega \| 100(26 \text{ }\Omega) = 2.3 \text{ k}\Omega$$

g. Check Z_o:

$$Z_o = R_C = 4.1 \text{ k}\Omega$$

Part 2. Computer Analysis of Design

PSpice Simulation 19-1

The common-emitter circuit shown was designed given the stated specifications in the "Resumé of Theory" section. A collector current of 1 milliampere was assumed in the determination of the circuit component values. A Vsignal of 15 mV peak voltage applied corresponds closely to a 10 mV rms signal as used in your laboratory experiment.

PSpice Simulation 19-1: Design of Common Emitter Amplifier

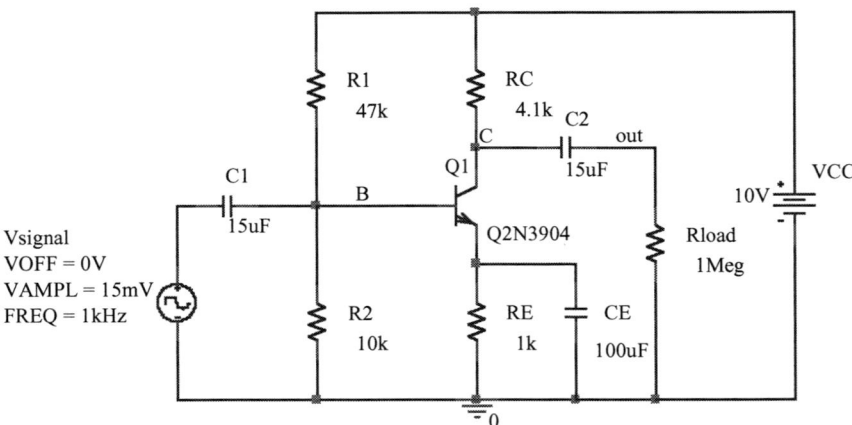

To start the checking procedure of the design, run a bias point analysis and answer the following questions:

1. What are the base and collector currents?

2. Are they reasonably close to their assumed values?

3. Calculate the β of the transistor.

4. What is the collector-to-emitter voltage?

5. Does it allow for a 3 V_{p-p} swing?

Determining the A_v, the input and the output impedances

Perform a Time Domain (transient) analysis of the above circuit of 20 milliseconds duration. To obtain the voltage gain:

1. Plot the ratio of RMS(V(Vout))/RMS(V(Vsignal: +)).

2. Does it meet the minimum value desired?

3. If not, suggest a procedure to obtain its desired value.

4. Next, plot the ratio of RMS(V(Vsignal))/RMS(I(C1)). This ratio is equal to the input impedance of the amplifier.

5. Does it meet the minimum value desired?

6. If not, suggest a procedure to obtain its desired value.

7. We shall next obtain the output impedance of the amplifier. Modify the above circuit as shown below.

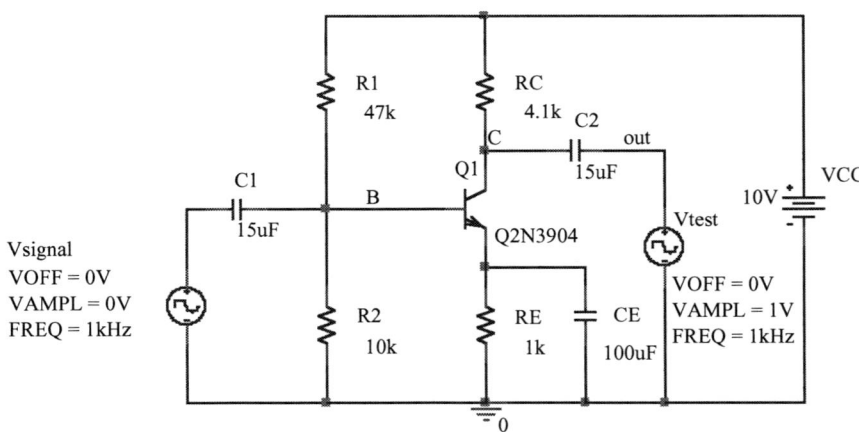

PSpice Simulation 19-1: Design of Common Emitter Amplifier

8. Run a Time Domain (transient) analysis of 20 milliseconds duration and obtain a plot of RMS(V(Vtest: +))/RMS(I(C2)). This ratio is equal to the output impedance of the amplifier.

Exp. 19 / Procedure

9. Is it below the maximum value desired?

10. If not, suggest a procedure to obtain its desired value.

Part 3. Build and Test CE Circuit

a. Build the CE amplifier circuit of Fig. 19.1 using the capacitors, resistors, and transistor from the design in Part 1, and analysis in Part 2.

b. Set $V_{CC} = 10$ V. Measure and record DC voltages.

V_B (measured) = _____
V_E (measured) = _____
V_C (measured) = _____

Calculate the value of $I_C = I_E$.

$I_C = I_E =$ _____

Calculate the dynamic resistance, r_e.

$r_e =$ _____

c. Apply an AC input, $V_{sig} = 10$ mV, rms at $f = 1$ kHz (or adjust value for maximum undistorted load voltage as observed using a scope). Measure and record AC voltages.

$V_{sig} =$ _____
V_L (measured) = _____

Calculate A_v with load resistor connected.

$$A_v = \frac{V_L}{V_{sig}}$$

$A_v =$ _____

d. Connect a measurement resistor, $R_x = 3$ kΩ, in series with input V_{sig}. Using a DMM, measure and record V_{sig} and V_i (from base to ground).

$V_{sig} =$ _____

V_i (measured) = _____

Calculate Z_i.

$$Z_i = \frac{R_x V_i}{V_{sig} - V_i} \; \Omega$$

$Z_i =$ _____

Remove resistor R_x.

e. Remove load resistor R_L. (Readjust V_{sig} if waveform seen using scope is distorted.) Measure unloaded AC output voltage V_o.

V_o (measured) = _____

Calculate AC output impedance (using V_L in Part 3(c)).

using: $\quad V_L = \dfrac{R_L}{R_L + Z_o} V_o$

from which: $\quad Z_o = \dfrac{V_o - V_L}{V_L} R_L$

$Z_o =$ _____

f. Provide a summary of original design specs and actual specs determined by measurement.

Provide a comparison to show whether the design procedure was successful. Indicate any possible factors if design results are not fully satisfactory.

Name _____
Date _____
Instructor _____

EXPERIMENT 20

Common-Source Transistor Amplifiers

OBJECTIVES

1. To measure DC and AC voltages in a common-source amplifier.
2. To obtain measured values of voltage amplification (A_v), input impedance (Z_i), and output impedance (Z_o).

EQUIPMENT REQUIRED

Instruments

Oscilloscope
DMM
Function generator
DC supply

Components

Resistors

(1) 510-Ω
(1) 1-kΩ
(1) 2.4-kΩ
(1) 10-kΩ
(2) 1-MΩ

Capacitors

(2) 15-μF
(1) 100-μF

Transistors

(1) 2N3823, or equivalent

EQUIPMENT ISSUED

Item	Laboratory serial no.
DC power supply	
Function generator	
Oscilloscope	
DMM	

RÉSUMÉ OF THEORY

The DC bias of a JFET is determined by the device transfer characteristic (V_P and I_{DSS}) and the DC self-bias determined by the source resistor. The AC voltage gain at this DC bias point is then dependent on the device parameters (g_m or g_{fs}) and circuit drain resistance.

AC Voltage Gain: The voltage gain of the amplifier shown in Fig. 20.1 is calculated from

$$A_v = \frac{V_o}{V_i} = -g_m R_D \qquad [\,= -g_m(R_D \| R_L)\,] \qquad (20.1)$$

where

$$g_m = g_{m0}(1 - V_{GSQ}/V_P) \quad \text{with } g_{m0} = \frac{2 I_{DSS}}{|V_p|} \qquad (20.2)$$

AC Input Impedance: The AC input impedance is

$$Z_i = R_G \qquad (20.3)$$

AC Output Impedance: The AC output impedance is

$$Z_o = R_D \qquad (20.4)$$

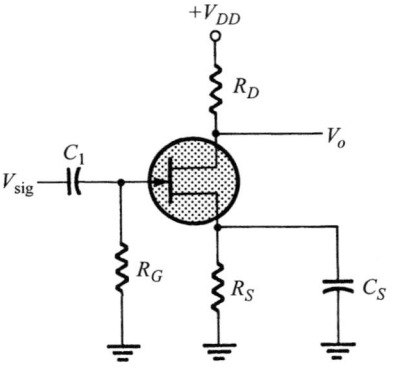

Figure 20-1

Exp. 20 / Procedure

PROCEDURE

Part 1. Measurement of I_{DSS} and V_P

Use a characteristic curve tracer to determine the values of I_{DSS} and V_P, if available. Otherwise, use the following steps to obtain these values.

    a. Construct the circuit of Fig. 20.1 with V_{DD} = +20 V, R_G = 1 MΩ, R_D = 510 Ω, and R_S = 0 Ω. Measure and record.

V_D (measured) = _____

Calculate the value of drain current, I_D.

$$I_D = \frac{V_{DD} - V_D}{R_D}$$

I_D (calculated) = _____

Since this is the drain current at V_{GS} = 0 V

$I_{DSS} = I_D$ = _____

(using the value of I_D just calculated).

    b. Now connect R_S = 1 kΩ. Measure and record the values of

V_{GS} (measured) = _____
V_D (measured) = _____

using the measured values just obtained:

Calculate V_P as follows.

First: $\quad I_D = \dfrac{V_{DD} - V_D}{R_D}$

I_D (calculated) = _____

Second: $$V_P = \frac{V_{GS}}{1 - \sqrt{\dfrac{I_D}{I_{DSS}}}}$$

V_P (calculated) = _____

Part 2. DC Bias of Common-Source Circuit

a. Calculate the DC bias expected in the circuit of Fig. 20.2, using I_{DSS} and V_P obtained in Part 1.

Draw graphs of the equations

$$I_D = I_{DSS}\left(1 - \frac{V_{GS}}{V_P}\right)^2 \text{ and } V_{GS} = -I_D R_S$$

to graphically obtain the equation intersection.
Or: use a computer or programmable calculator to solve the simultaneous equations.

The calculated DC bias values are:

V_{GS} (calculated) = _____
I_D (calculated) = _____

using

$$V_D = V_{DD} - I_D R_D$$

V_D (calculated) = _____

b. Build the circuit of Fig. 20.2 using $R_G = 1\ M\Omega$, $R_S = 510\ \Omega$, and $R_D = 2.4\ k\Omega$. Set $V_{DD} = +20\ V$.

c. Measure the DC bias voltages.

V_G (measured) = _____
V_S (measured) = _____
V_D (measured) = _____
V_{GS} (measured) = _____

Calculate the value of I_D under DC bias conditions.

$$I_D = \frac{V_S}{R_S}$$

I_D = _____

Compare the DC bias values calculated in Part 2(**a**) with those measured in Part 2(**c**).

Part 3. AC Voltage Gain of Common-Source Amplifier

a. Calculate the voltage gain of the common-source amplifier of Fig. 20.2.

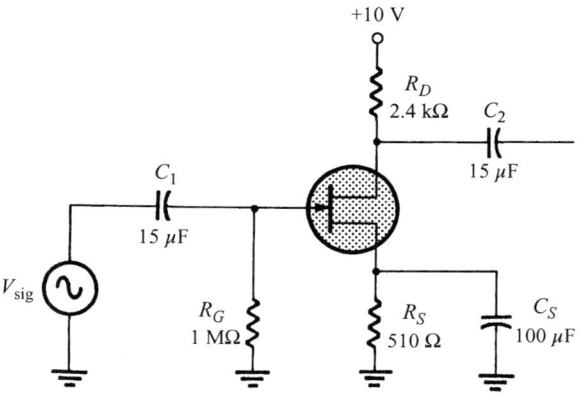

Figure 20-2

$$A_v = -g_m R_D$$

$$\text{with } g_m = \left(\frac{2I_{DSS}}{|V_P|}\right)\left(1 - \frac{V_{GS}}{V_P}\right)^2$$

Use V_P and I_{DSS} from Part 1, and V_{GS} calculated in Part 2.

A_v (calculated) = _____

b. Connect the input of $V_{sig} = 100$ mV at 1 kHz. Measure and record using the DMM:

V_o (measured) = _____

Calculate the voltage gain of the amplifier.

$$A_v = \frac{V_o}{V_{sig}}$$

A_v = _____

Part 4. Input and Output Impedance Measurements

a. The input impedance is

$$Z_i = R_G$$

Z_i (expected) = _____

b. The output impedance is

$$Z_o = R_D$$

Z_o (expected) = _____

c. Connect a 1-MΩ resistor, R_x, in series with the input signal, V_{sig} = 100 mV, rms at f = 100 Hz. Measure V_i.

V_i (measured) = _____

Determine the input impedance using

$$Z_i = \frac{Z_i}{V_{sig} - V_i} R_x$$

Z_i (calculated) = _____

Remove the measurement resistor, R_x.

d. Measure V_o.

V_o (measured) = _____

Connect load R_L = 10 kΩ. Measure voltage across load, V_L.

V_L (measured) = _____

Determine the AC output impedance using

$$Z_o = \frac{V_o - V_L}{V_L} R_L$$

Z_o (calculated) = _____

Compare the input impedance calculated in Part 4(**a**) with that determined from measurements in Part 4(**c**).

Compare the output impedance calculated in Part 4(**b**) with that determined from measurements in Part 4(**d**).

Part 5. Computer Exercise

PSpice Simulation 20-1

The circuit shown is that of Fig. 20.2.

PSpice Simulation 20-1: Common Source amplifier

Recall from Experiment 12 that for the J2N4393, I_{DSS} was 15.86 mA and the absolute value of V_p was 1.5 volts.

1. To begin the analysis of this amplifier, run a bias point analysis and obtain the DC voltages and currents of this circuit.

2. Compare their values with those obtained experimentally.

3. Calculate the transconductance of the JFET.

4. Calculate the theoretical value of the voltage gain.

5. Perform a Time Domain (transient) analysis of 20 milliseconds duration.

6. Obtain the ratio of RMS(V(OUT))/RMS(V(SIGNAL)). This ratio is equal to the voltage gain of the amplifier.

7. How does its value compare to its theoretical value and that obtained experimentally?

8. Obtain the ratio RMS(V(SIGNAL))/RMS(I(C1)). This ratio is equal to the input impedance of the amplifier.

9. How does its value compare to its theoretical value and that obtained experimentally?

10. To obtain the output impedance, modify the circuit as shown.

PSpice Simulation 20-1: Common Source amplifier

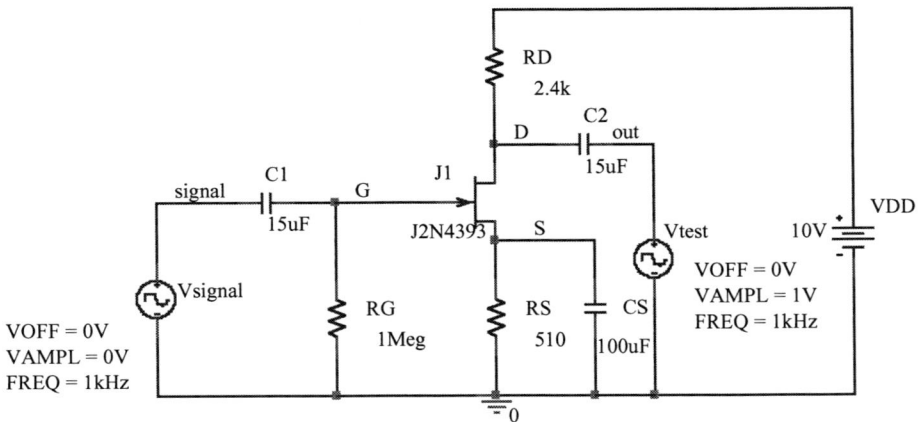

11. Obtain the ratio RMS(V(Vtest: +))/RMS(I(C2)). This ratio is equal to the output impedance of the amplifier.

12. Compare its value to its theoretical value and that obtained experimentally.

13. Which of the three, the voltage gain, the input impedance, or the output impedance, is dependent upon the transconductance?

14. Using the first of the two circuits in this simulation, change resistance RD to 4 kΩ and run a Time Domain (transient) analysis of 20 milliseconds duration to determine a possible new value for the voltage gain.

15. Explain the difference between the newly obtained voltage gain and its previous value.

Name _____
Date _____
Instructor _____

EXPERIMENT 21

Multistage Amplifiers: *RC* Coupling

OBJECTIVES

1. To measure DC and AC voltages in a multistage FET amplifier.
2. To obtain measured values of voltage amplification (A_v), input impedance (Z_i), and output impedance (Z_o).

EQUIPMENT REQUIRED

Instruments

Oscilloscope
DMM
Function generator
DC supply

Components

Resistors

(2) 510-Ω
(1) 1-kΩ
(2) 2.4-kΩ
(1) 10-kΩ
(3) 1-MΩ

Capacitors

(3) 15-μF
(2) 100-μF

Transistors

(2) 2N3823, or equivalent

EQUIPMENT ISSUED

Item	Laboratory serial no.
DC power supply	
Function generator	
Oscilloscope	
DMM	

RÉSUMÉ OF THEORY

The DC bias of a JFET is determined by the device transfer characteristic (V_P and I_{DSS}) and the external circuit connected to it. The AC voltage gain at this DC bias point is dependent on the device parameters (g_m or g_{fs}) and circuit drain resistance.

AC Voltage Gain: The voltage gain of an amplifier stage as shown in Fig. 21.1 can be calculated from

$$A_v = \frac{V_o}{V_i} = -g_m R_D = -g_m(R_D \| R_L) \qquad (21.1)$$

where

$$g_m = g_{m0}\left(1 - \frac{V_{GSQ}}{V_P}\right) \quad \text{with } g_{m0} = \frac{2I_{DSS}}{|V_p|} \qquad (21.2)$$

AC Input Impedance: The AC input impedance is

$$Z_i = R_G \qquad (21.3)$$

AC Output Impedance: The AC output impedance is

$$Z_o = R_D \qquad (21.4)$$

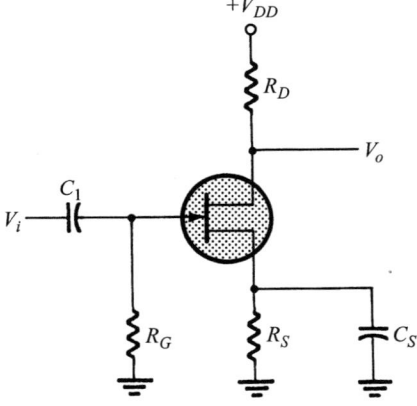

Figure 21-1

PROCEDURE

Part 1. Measurement of I_{DSS} and V_P

It is necessary to obtain values of I_{DSS} and V_P for both Q_1 and Q_2. Use a characteristic curve tracer, if available, to determine the values of I_{DSS} and V_P. Obtain readings at $V_{DS} = +10$ V.

For Q_1:

I_{DSS} = _____
V_P = _____

For Q_2:

I_{DSS} = _____
V_P = _____

Go on to Part 2.

If no curve tracer is available, use the following steps to obtain the above values.

a. Construct the circuit of Fig. 21.2 with $R_D = 510\ \Omega$ but with $R_S = 0\ \Omega$. Measure and record.

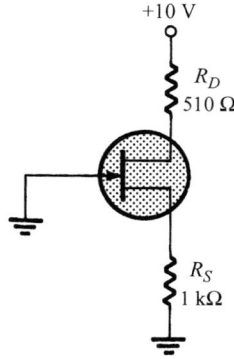

Figure 21-2

V_D (measured) = _____

Calculate the value of drain current, I_D.

$$I_D = \frac{V_{DD} - V_D}{R_D}$$

I_D (calculated) = _____

Since this is the drain current at $V_{GS} = 0$ V

$I_{DSS}(Q_1) = I_D =$ _____

(using the value of I_D just calculated).

Replace Q_1 and repeat the measurement with Q_2.

V_D (measured) = _____

Calculate the value of drain current, I_D.

$$I_D = \frac{V_{DD} - V_D}{R_D}$$

I_D (calculated) = _____

Since this is the drain current at $V_{GS} = 0$ V

$I_{DSS}(Q_2) = I_D =$ _____

(using the value of I_D just calculated).

Exp. 21 / Procedure

b. Now connect $R_S = 1\ \text{k}\Omega$. Measure and record the values of

V_{GS} (measured) = _____
V_D (measured) = _____

Using the measured values just obtained, calculate V_P as follows.

$$I_D = \frac{V_{DD} - V_D}{R_D}$$

I_D (calculated) = _____

$$V_P = \frac{V_{GS}}{1 - \sqrt{\dfrac{I_D}{I_{DSS}}}}$$

$V_P(Q_2)$ (calculated) = _____

Replace transistor Q_2 and repeat Part 1(**b**) measurements.

V_{GS} (measured) = _____
V_D (measured) = _____

Using the measured values just obtained, calculate V_P as follows.

$$I_D = \frac{V_{DD} - V_D}{R_D}$$

I_D (calculated) = _____

$$V_P = \frac{V_{GS}}{1 - \sqrt{\frac{I_D}{I_{DSS}}}}$$

$V_P(Q_1)$ (calculated) = _____

Part 2. DC Bias of Common-Source Circuit

a. Calculate the DC bias expected in the circuit of Fig. 21.3, using I_{DSS} and V_P obtained in Part 1 for each transistor.

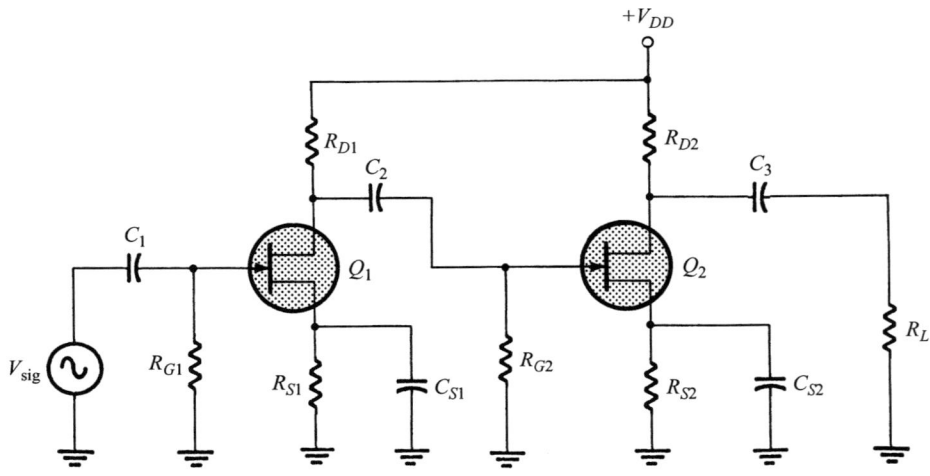

Figure 21-3

Draw graphs of the equations

$$I_D = I_{DSS}\left(1 - \frac{V_{GS}}{V_P}\right)^2 \text{ and } V_{GS} = -I_D R_S$$

to obtain the equations' intersection.

Or use a computer or programmable calculator to solve the simultaneous equations.

The calculated DC bias values are:

V_{GS1} (calculated) = _____

Exp. 21 / Procedure

I_{D_1} (calculated) = _____

using

$$V_{D1} = V_{DD} - I_{D_1} R_{D_1}$$

V_{D_1} (calculated) = _____

The calculated DC bias values are:

V_{GS_2} (calculated) = _____
I_{D_2} (calculated) = _____

using

$$V_{D_2} = V_{DD} - I_{D_2} R_{D_2}$$

V_{D_2} (calculated) = _____

b. Build the circuit of Fig. 21.3 using $R_{G_1} = R_{G_2} = 1$ MΩ, $R_{S_1} = R_{S_2} = 510$ Ω, and $R_{D_1} = R_{D_2} = 2.4$ kΩ. $R_L = 10$ kΩ. Set $V_{DD} = +20$ V.

c. Measure the DC bias voltages.

V_{G_1} (measured) = _____
V_{S_1} (measured) = _____
V_{D_1} (measured) = _____
V_{GS_1} (measured) = _____

Calculate the value of I_{D_1} under DC bias conditions (using nominal resistor values).

$$I_{D_1} = \frac{V_{S_1}}{R_{S_1}}$$

I_{D_1} = _____
V_{G_2} (measured) = _____
V_{S_2} (measured) = _____
V_{D_2} (measured) = _____
V_{GS_2} (measured) = _____

Calculate the value of I_{D_2} under DC bias conditions.

$$I_{D_2} = \frac{V_{S_2}}{R_{S_2}}$$

$I_{D_2} = $ _____

Compare the DC bias values calculated in Part 2(**a**) with those measured in Part 2(**c**).

Part 3. AC Voltage Gain of Amplifier

a. Calculate the voltage gain of the common-source amplifier of Fig. 21.3.

For stage 2:

$$A_{v_2} = -g_m(R_{D_2} \| R_L)$$

$$\text{with } g_m(Q_2) = \frac{2I_{DSS}(Q_2)}{|V_p(Q_2)|}\left(1 - \frac{V_{GS_2}}{V_p(Q_2)}\right)$$

Using $V_P(Q_2)$, $I_{DSS}(Q_2)$ from Part 1, and V_{GS_2} calculated in Part 2

A_{V_2} (calculated) = _____

For stage 1:

$$A_{v_1} = -g_{m_1}(R_{D1} \| Z_{i_2})$$

$$\text{with } g_m(Q_1) = \frac{2I_{DSS}(Q_1)}{|V_p(Q_1)|}\left(1 - \frac{V_{GS_1}}{V_p(Q_1)}\right)$$

Using $V_P(Q_1)$, $I_{DSS}(Q_1)$ from Part 1, and V_{GS_1} calculated in Part 2

A_{V_1} (calculated) = _____

Exp. 21 / Procedure

Calculate the overall amplifier gain:

$$A_v = A_{v_1} \times A_{v_2}$$

A_v (calculated) = _____

b. Connect the input of $V_{sig} = 10$ mV, rms at $f = 1$ kHz. Use the oscilloscope to obtain an undistorted output voltage, adjusting V_{sig} if necessary. Measure and record:

V_{sig} (measured) = _____
V_L (measured) = _____

Calculate the voltage gain of the overall amplifier:

$$A_v = \frac{V_L}{V_{sig}}$$

A_v = _____

Measure and record:

V_{o_1} (measured) = _____

Calculate the gain of each stage:

$$A_{v_1} = \frac{V_{o_1}}{V_{sig}}$$

A_{v_1} (calculated) = _____

$$A_{v_2} = \frac{V_L}{V_{o1}}$$

A_{v_2} (calculated) = _____

Part 4. Input and Output Impedance Measurements

a. The input impedance is

$$Z_i = R_{G_1}$$

Z_i = _____

b. The output impedance is

$$Z_o = R_{D_2}$$

Z_o = _____

c. Connect a 1-MΩ resistor, R_x, in series with the input signal, $V_{\text{sig}} = 10$ mV, rms at $f = 100$ Hz. Measure V_{i_1}.

V_{i_1} (measured) = _____

Calculate the input impedance using

$$Z_i = \frac{V_{i_1}}{V_{\text{sig}} - V_{i_1}} R_x$$

Z_i = _____

Remove the measurement resistor, R_x.

d. Measure V_L.

V_L (measured) = _____

Disconnect load $R_L = 10$ kΩ. Measure output voltage, V_o.

V_o (measured) = _____

Calculate the AC output impedance using

$$Z_o = \frac{V_o - V_{i_1}}{V_L} R_L$$

Z_o = _____

Compare the input impedance calculated in Part 4(**a**) with that determined from measurements in Part 4(**c**).

Compare the output impedance calculated in Part 4(**b**) with that determined from measurements in Part 4(**d**).

Part 5. Computer Exercise

PSpice Simulation 21-1

The circuit shown is that of Fig. 21.3. It is an RC-coupled multistage amplifier. Note that the JFETs used are the same as in the previous experiment. Thus IDSS is 15.86 mA and the absolute voltage of V_p is 1.5 volts.

Exp. 21 / Multistage Amplifiers: *RC* Coupling

PSpice Simulation 21-1: Multistage Common Source Circuit

[Circuit schematic: Two-stage common source JFET amplifier using J2N4393 transistors. Vsignal (VOFF = 0V, VAMPL = 10mV, FREQ = 1kHz) connects through C1 (15uF) to the gate of J1. RG1 = 1Meg to ground, RS1 = 510 with CS1 = 100uF bypass, RD1 = 2.4k to +10V supply. Output through C2 (15uF) to node D1G2 at gate of J2. RG2 = 1Meg, RS2 = 510 with CS2 = 100uF bypass, RD2 = 1k. Output through C3 (15uF) to Rload = 1Meg.]

1. Run a bias point analysis and obtain the DC voltages and currents of the two stages.

2. Compare their values with those obtained experimentally.

3. Calculate the tranconductances for the two JFETs.

4. Calculate the theoretical voltage gain of each stage.

5. Perform a Time Domain (transient) analysis of 20 milliseconds duration.

6. Obtain the voltage gain of the first stage. Its expression is: RMS(V(D1G2))/RMS(V(IN)).

7. Obtain the voltage gain of the second stage. Its expression is: RMS(V(OUT))/RMS(V(D1G2)).

8. Compare the two voltage gains.

9. Why are their values different?

10. What is the value of their product?

11. Obtain the overall value of the voltage gain of the two stages. Its expression is: RMS(V(OUT))/RMS(V(IN)).

12. Compare this value with that calculated in step 9.

13. Compare the voltage gains of stage 1 and stage 2 and the overall voltage gain with their values obtained experimentally.

14. Are they within 10% of agreement?

15. If not, explain a possible reason for their differences.

16. How can you prove that the two stages are biased independently of each other?

17. Obtain the ratio of RMS(V(IN))/RMS(I(C1)). This ratio is equal to the input impedance of the amplifier.

18. Compare its value to its theoretical and experimentally obtained values.

19. To obtain the output impedance next, modify the circuit as shown.

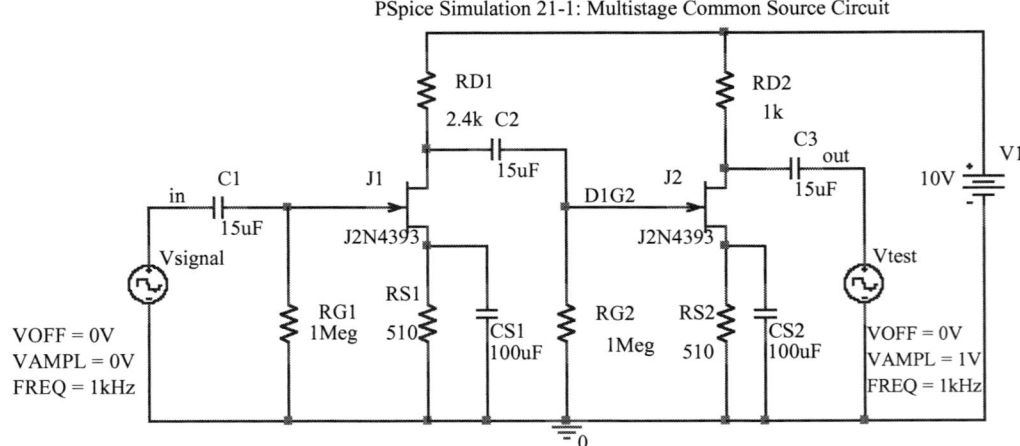

20. Obtain the ratio of RMS(V(OUT))/RMS(I(C3)). This ratio is equal to the output impedance of the amplifier.

21. Compare its value to its theoretical and experimental value.

22. Explain any possible difference in their values.

Name _____
Date _____
Instructor _____

EXPERIMENT 22
CMOS Circuits

OBJECTIVE

To measure DC and AC operation in CMOS circuits.

EQUIPMENT REQUIRED

Instruments

Oscilloscope
DMM
Function generator
DC supply

Components

Integrated circuits

(1) 74HC02, or 14002
(1) 74HC04, or 14004

EQUIPMENT ISSUED

Item	Laboratory serial no.
DC power supply	
Function generator	
Oscilloscope	
DMM	

RÉSUMÉ OF THEORY

A CMOS circuit can be built using two opposite type MOSFET devices as shown in Fig. 22.1. Digital inputs are 0 V and 5 V. For an input of 0 V the n-type enhancement MOSFET (nMOS) device is turned *off*, while the p-type enhancement MOSFET (pMOS) device is turned *on*, as shown in Fig. 22.2a. An input of 5 V will drive the pMOS device *off* and the nMOS device *on* with the output then near 0 V, as shown in Fig. 22.2b.

A CMOS gate with two inputs is shown in Fig. 22.3. Each input is connected to a pair of pMOS and nMOS transistors. The operation for various inputs of 0 V and 5 V is summarized in Figure 22.3. When both inputs are 0 V, the two pMOS devices are *on*, both nMOS devices are *off*, and the output is 5 V. When both inputs are 0 V, or even one input is 0 V, one nMOS device is *on*, one pMOS device is *off*, and the output is near 0 V.

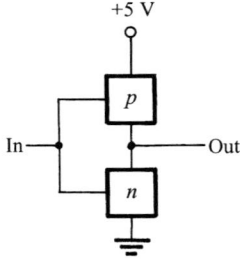

Figure 22-1

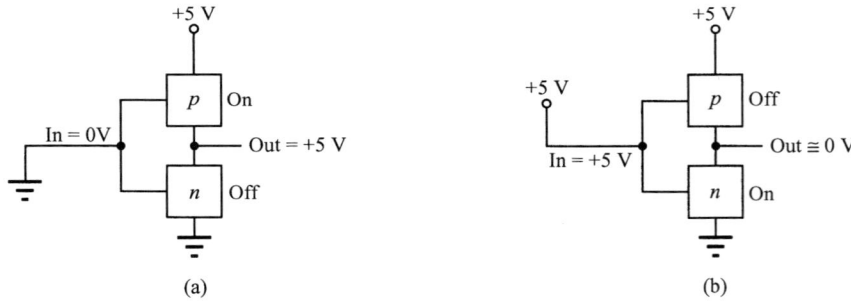

Figure 22-2

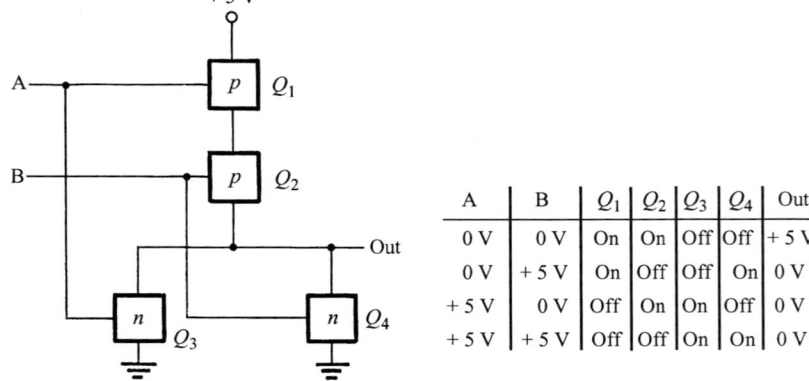

A	B	Q_1	Q_2	Q_3	Q_4	Out
0 V	0 V	On	On	Off	Off	+5 V
0 V	+5 V	On	Off	Off	On	0 V
+5 V	0 V	Off	On	On	Off	0 V
+5 V	+5 V	Off	Off	On	On	0 V

Figure 22-3

PROCEDURE

Part 1. CMOS Inverter Circuit

a. Construct the CMOS inverter circuit shown in Fig. 22.1.

b. For the CMOS inverter circuit of Fig. 22.1 determine the output voltage for inputs of 0 V and 5 V and record in Table 22.1.

TABLE 22.1

IN	OUT
0 V	
5 V	

c. Connect 5 V to an inverter IC such as the 74HC04 or 14004. Apply inputs of 0 V and 5 V, recording outputs in Table 22.2.

TABLE 22.2

IN	OUT
0 V	
5 V	

d. Apply a clock signal ($f = 10$ kHz) as input and record the input and output waveforms viewed on an oscilloscope in Fig. 22.4.

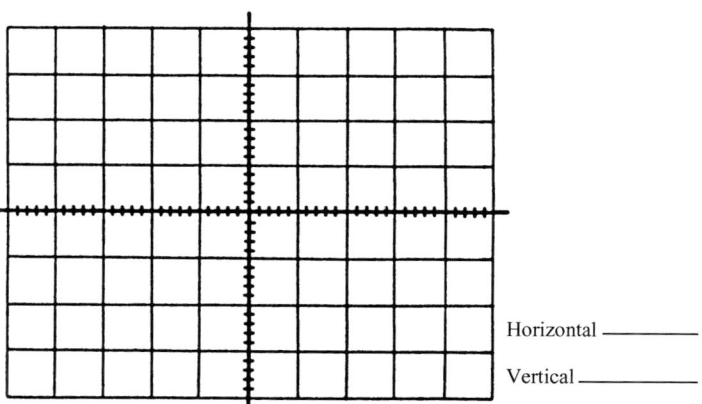

Horizontal _____
Vertical _____

Figure 22-4

Part 2. CMOS Gate

a. Connect power to a CMOS IC such as a 74HC02 or 14002 as shown in Fig. 22.5. Apply inputs of 0 V and 5 V and record output in Table 22.3.

Figure 22-5

TABLE 22.3

A	B	OUTPUT
0 V	0 V	
0 V	5 V	
5 V	0 V	
5 V	5 V	

b. Connect 0 V to one input and a digital clock to the other. Observe and record the output waveform in Fig. 22.6a.

c. Connect 5 V to one input and a digital clock to the other. Observe and record the output waveform in Fig. 22.6b.

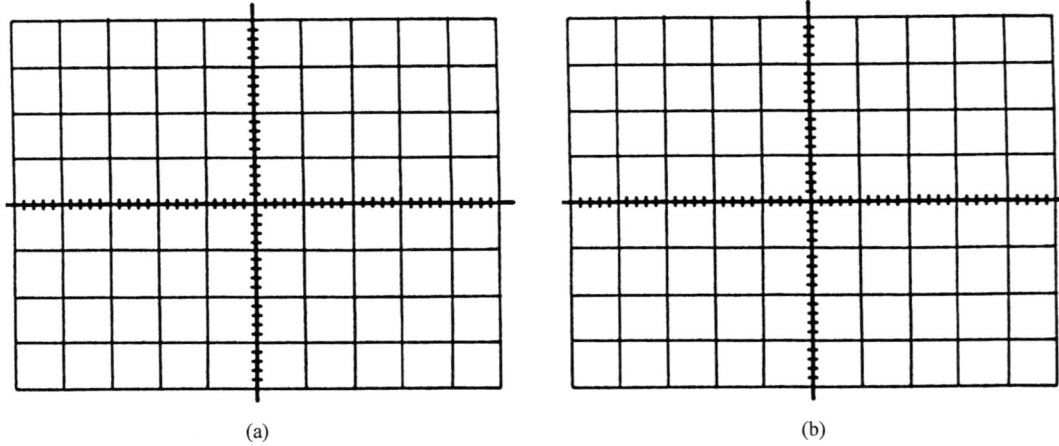

Figure 22-6

Part 3. CMOS Input-Output Characteristic

a. Using the CMOS inverter circuit (74HC040) with variable input as shown in Fig. 22.7, complete Table 22.4.

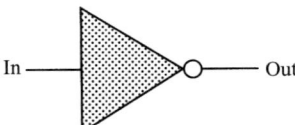

Figure 22-7

TABLE 22.4

IN	0.0	0.2	0.4	0.6	0.8	1.0	1.2	1.4	1.6	1.8	2.0	2.2
OUT												

IN	2.4	2.6	2.8	3.0	3.2	3.4	3.6	3.8	4.0	4.2	4.4	4.6	4.8	5.0
OUT														

b. Plot the data from Table 22.4 in Fig. 22.8.

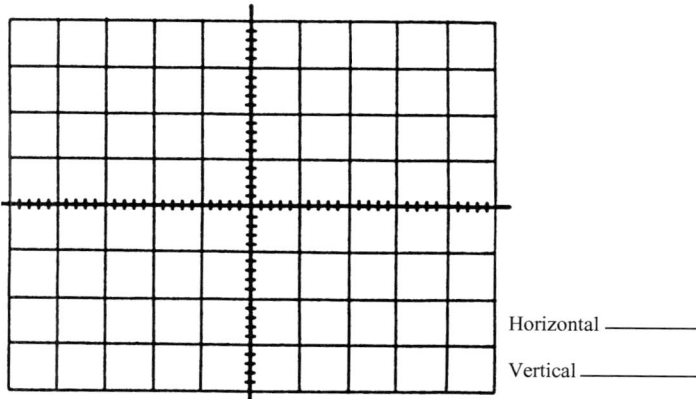

Horizontal ———

Vertical ———

Figure 22-8

Part 4. Computer Exercise

PSpice Simulation 22-1

The gate shown in the circuit has the same shape as the gate shown in Fig. 22.5. Thus, their logic functions are the same, namely, that of an OR gate. Terminals 1 and 2 are the inputs; terminal 3 is the output.

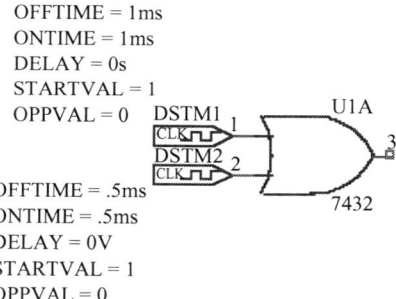

PSpice Simulation 22-1: CMOS Gate Logic

OFFTIME = 1ms
ONTIME = 1ms
DELAY = 0s
STARTVAL = 1
OPPVAL = 0

OFFTIME = .5ms
ONTIME = .5ms
DELAY = 0V
STARTVAL = 1
OPPVAL = 0

DSTM1 and DSTM2 are clocks connected to the input terminals of the gate and with their parameters set as shown. The logic states assumed by the clocks correspond to those listed for the A and B terminals in Table 22.3. To obtain the OUTPUT, run a time (transient) analysis of 10 milliseconds duration. On the Probe plot, obtain the traces of the two input terminals and the output terminal. Activate one of the cursors and run it along the time axis. For each position of the cursor, the logic states for the three terminals will be shown on the left margin of the plots.

1. Compare the results obtained from this simulation with those obtained experimentally.

2. In the logical analysis of this gate, is it important to know the internal structure of the 7432 gate?

To obtain the voltage levels at the terminals rather than their logical states, modify the above circuit by adding VPLOT1 devices to the gate as shown. A 5 V level at the input terminals, or a 3.5 V level at the output terminal corresponds to a logical 1 state. Zero volts at any of the three terminals corresponds to a logical 0 state.

PSpice Simulation 22-1: CMOS Gate Logic

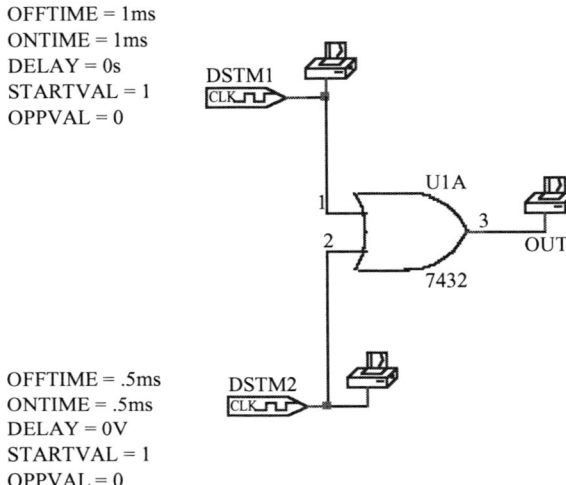

Use three Probe plots of these voltages to increase the clarity of the results.

From the results obtained:

1. Classify this logic gate from the Probe plot data.

2. How does its logic compare to that of the CMOS gate used in the laboratory experiment?

3. Do the voltage levels at the terminals correspond to those of the laboratory experiment?

4. Why is the voltage at the output terminal less than that of either input terminal?

Darlington and Cascode Amplifier Circuits

OBJECTIVE

To calculate and measure DC and AC voltages in Darlington and cascode connection circuits.

EQUIPMENT REQUIRED

Instruments

Oscilloscope
DMM
Function generator
DC supply

Components

Resistors

(1) 100-Ω
(1) 51-Ω, 1-W
(1) 1-kΩ
(1) 1.8-kΩ
(1) 4.7-kΩ
(1) 5.6-kΩ
(1) 6.8-kΩ
(1) 50-kΩ pot
(1) 100-kΩ

Capacitors

(1) 0.001-µF
(4) 10-µF

Transistors

(2) 2N3904 (or equivalent general purpose npn)
(1) TIP120 (npn Darlington)

EQUIPMENT ISSUED

Item	Laboratory serial no.
DC power supply	
Function generator	
Oscilloscope	
DMM	

RÉSUMÉ OF THEORY

Darlington Circuit: A Darlington connection (as shown in Fig. 23.1) provides a pair of BJT transistors in a single IC package with effective beta (β_D) equal to the product of the individual transistor betas.

$$\beta_D = \beta_1 \beta_2 \tag{23.1}$$

The Darlington emitter-follower has a higher input impedance than that of an emitter-follower. The Darlington emitter-follower input impedance is

$$Z_i = R_B \| (\beta_D R_E) \tag{23.2}$$

The output impedance of the Darlington emitter-follower is

$$Z_o = r_e \tag{23.3}$$

The voltage gain of a Darlington emitter-follower circuit is

$$A_v = \frac{R_E}{(R_E + r_e)} \tag{23.4}$$

Cascode Circuit: A cascode circuit, as shown in Fig. 23.2, provides a common-emitter amplifier using Q_1 directly connected to a common-base amplifier using Q_2. The voltage gain of stage Q_1 is approximately 1, with the voltage V_{o1} being opposite in polarity to V_i.

$$A_{v_1} = -1 \tag{23.5}$$

The voltage gain of stage Q_2 is noninverted and of magnitude

$$A_{v_2} = \frac{R_C}{r_{e_2}} \tag{23.6}$$

The overall gain is:

$$A_v = A_{v_1} A_{v_2} = -\frac{R_C}{r_{e_2}} \tag{23.7}$$

PROCEDURE

Part 1. Darlington Emitter-Follower Circuit

a. For the circuit of Fig. 23.1 calculate the DC bias voltages and currents.

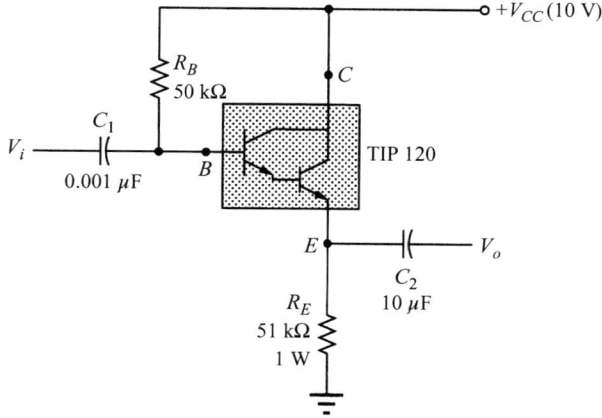

Figure 23-1

V_B (calculated) = _____
V_E (calculated) = _____

Calculate the theoretical values of voltage gain and input and output impedance.

A_v (calculated) = _____
Z_i (calculated) = _____
Z_o (calculated) = _____

b. Construct the Darlington circuit of Fig. 23.1. Adjust the 50-kΩ potentiometer (R_B) to provide an emitter voltage, V_E = 5 V. Using a DMM, measure and record the DC bias values:

V_B (measured) = _____
V_E (measured) = _____

Calculate the base and emitter DC currents:

I_B (calculated) = _____
I_E (calculated) = _____

Calculate the value of transistor beta at this Q-point:

β_D (calculated) = _____

c. Apply an input signal V_{sig} = 1 V, peak at f = 10 kHz. Using the oscilloscope, observe and record the output voltage to assure that the signal is not clipped or distorted. (Reduce the input signal amplitude if necessary.)

V_i (measured) = _____
V_o (measured) = _____

Calculate and record the AC voltage gain:

$A_v = V_o/V_i =$ _____

Part 2. Darlington Input and Output Impedance

a. Calculate the input impedance:

Z_i (calculated) = _____

Calculate the circuit output impedance:

Z_o (calculated) = _____

b. Connect a measurement resistor, $R_x = 100$ kΩ, in series with V_{sig}. Measure and record input voltage, V_i.

V_i (measured) = _____

Calculate the circuit input impedance using

$$Z_i = \frac{V_i}{V_{sig} + V_i} R_x$$

Z_i (calculated) = _____

Remove measurement resistor, R_x.

c. Measure the output voltage, V_o, with no load connected.

V_o (measured) = _____

Connect a load resistor, $R_L = 100$ Ω. Measure and record the resulting output voltage:

V_o (measured) = V_L = _____

Calculate the output impedance using

$$Z_o = \frac{V_o - V_L}{V_L} R_L$$

Z_o (calculated) = _____

Compare the calculated and measured values of Z_i and Z_o.

Part 3. Cascode Amplifier

a. Calculate DC bias voltages and currents in the cascode amplifier of Fig. 23.2 (assuming base currents are much less than the voltage divider current).

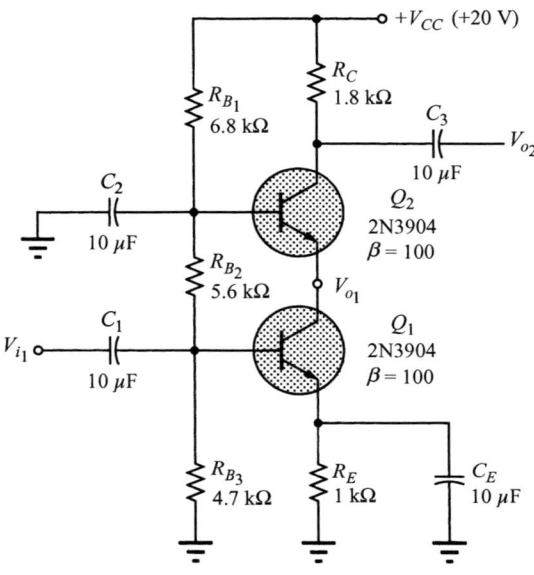

Figure 23-2

V_{B_1} (calculated) = _____
V_{E_1} (calculated) = _____
V_{C_1} (calculated) = _____
V_{B_2} (calculated) = _____
V_{E_2} (calculated) = _____
V_{C_2} (calculated) = _____

Calculate the DC bias emitter currents:

I_{E_1} (calculated) = _____
I_{E_2} (calculated) = _____

Calculate the transistor dynamic resistances:

r_{e_1} (calculated) = _____
r_{e_2} (calculated) = _____

b. Connect the cascode circuit of Fig. 23.2. Measure and record DC bias voltages.

V_{B_1} (measured) = _____
V_{E_1} (measured) = _____
V_{C_1} (measured) = _____
V_{B_2} (measured) = _____
V_{E_2} (measured) = _____
V_{C_2} (measured) = _____

Calculate the values of emitter current:

I_{E_1} = _____
I_{E_2} = _____

and the values of dynamic resistance:

r_{e_1} = _____
r_{e_2} = _____

c. Using Eqs. 23.5 and 23.6, calculate the AC voltage gain of each transistor stage:

A_{v_1} (calculated) = _____
A_{v_2} (calculated) = _____

d. Apply input signal, V_{sig} = 10 mV, peak at f = 10 kHz. Using the oscilloscope, observe the output waveform V_o to make sure that no signal distortion occurs. If the output is clipped or distorted, reduce the input signal until the clipping or distortion disappears.

Using the DMM, measure and record the AC signals.

V_i (measured) = _____
V_{o_1} (measured) = _____
V_{o_2} (measured) = _____

Calculate the measured voltage gains:

$A_{v_1} = V_{o_1}/V_i =$ _____
$A_{v2} = V_{o_2}/V_{o_1} =$ _____
$A_v = V_{o_2}/V_i =$ _____

Compare the measured voltage gains with those calculated in Parts 3(c) and 3(d).

e. Using the oscilloscope, observe and record waveforms for the input signal, V_i, output of stage 1, V_{o_1}, and output of stage 2, V_{o_2}. Show amplitude and phase relations clearly.

Part 4. Computer Exercises

PSpice Simulation 23-1

The Darlington emitter-follower circuit shown is that of Fig. 23.1.

PSpice Simulation 23-1: Darlington Emitter Follower

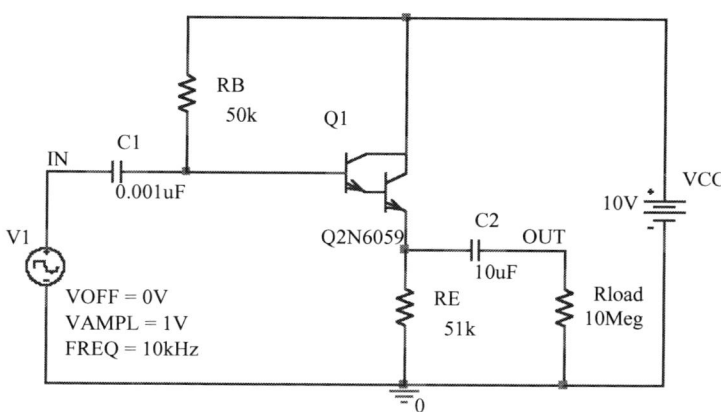

1. Perform a bias point analysis and obtain all the DC currents and voltages in the circuit.

2. From this data, calculate the dynamic resistance of the Darlington.

3. Perform a Time Domain (transient) analysis of 200 microseconds duration.

4. From this data, obtain Probe plots of the input voltage V(IN) versus the output voltage V(OUT).

5. What is the phase relationship between these two voltages?

6. Obtain the voltage gain of this amplifier.

7. Compare it to its theoretical value as calculated from Eq. 23.4.

8. Does its magnitude conform to that of an emitter-follower?

9. Obtain both the input and the output impedance using the procedure previously used in the PSpice simulations.

10. Compare these values to those obtained from experimental data.

PSpice Simulation 23-2

The next circuit in this simulation is that of the cascode amplifier of Fig. 23.2.

PSpice Simulation 23-2: Cascode Amplifier

1. Perform a bias point analysis and obtain all DC currents and voltages.

2. From this data, calculate the dynamic resistances of the two stages.

3. Compare their values with those obtained from your laboratory data.

4. Perform a Time Domain (transient) analysis of 200 microseconds duration.

5. Obtain Probe plots of the input voltage V(IN) and the output voltages V(OUT1) and V(OUT2).

6. Compare their relative amplitudes and phase relationship.

7. Obtain the Probe plots of the voltage gains of stage 1 and stage 2 and the overall voltage gain of the circuit.

8. Why are the gains of stages 1 and 2 so different in amplitude?

9. Compare the value of the three gains to those obtained from laboratory data.

EXPERIMENT 24

Current Source and Current Mirror Circuits

OBJECTIVE

To calculate and measure DC voltages in current source and current mirror circuits.

EQUIPMENT REQUIRED

Instruments

Oscilloscope
DMM
Function generator
DC supply

Components

Resistors

(1) 20-Ω
(1) 51-Ω
(1) 82-Ω
(1) 100-Ω
(1) 150-Ω
(2) 1.2-kΩ
(1) 3.6-kΩ
(1) 4.3-kΩ
(1) 5.1-kΩ
(1) 7.5-kΩ
(1) 10-kΩ

Transistors

(3) 2N3904, or equivalent npn transistor
(1) 2N3823, or equivalent JFET n-channel transistor

EQUIPMENT ISSUED

Item	Laboratory serial no.
DC power supply	
Function generator	
Oscilloscope	
DMM	

RÉSUMÉ OF THEORY

Current source and current mirror circuits are part of many types of linear integrated circuits. This experiment will build and test a few types of each circuit.

Current Source: Fig. 24.1 shows a simple form of current source using a JFET biased to operate at its drain-source saturation current. Regardless of the load R_L (within practical limits), the current through load R_L will be set by the JFET device:

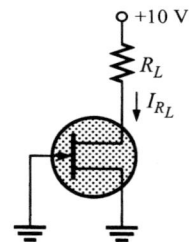

Figure 24-1

$$I_L = I_{DSS} \tag{24.1}$$

A BJT current source circuit is shown in Fig. 24.2. The base voltage is approximately set by

$$V_B = \frac{R_1}{R_1 + R_2}(-V_{EE})$$

The emitter voltage is then

$$V_E = V_B - 0.7 \text{ V}$$

with the emitter current then

$$I_{R_E} = \frac{V_E - V_{EE}}{R_E} = I_{R_L} \tag{24.2}$$

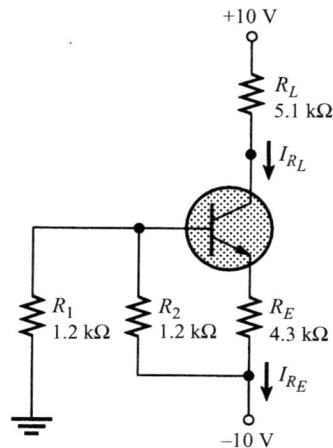

Figure 24-2

Current Mirror: The circuit of Fig. 24.3 is a current mirror, in which the current set through resistor R_x is mirrored through the load

$$I_x = \frac{V_{CC} - V_{BE}}{R_x} = I_{R_L} \tag{24.3}$$

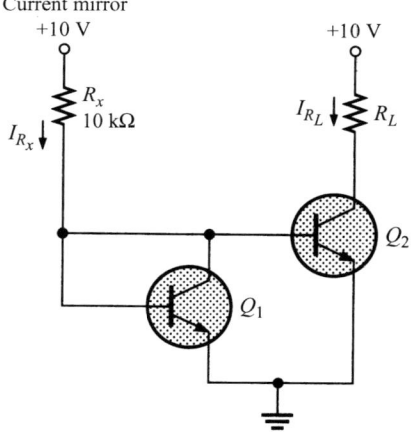

Figure 24-3

The circuit of Fig. 24.4 shows how a current mirror can provide the same current to a number of loads. The mirrored current set through resistor R_x and mirrored through both loads is

$$I_{R_x} = \frac{V_{CC} - V_{BE}}{R_X} = I_{R_2} = I_{R_3} \tag{24.4}$$

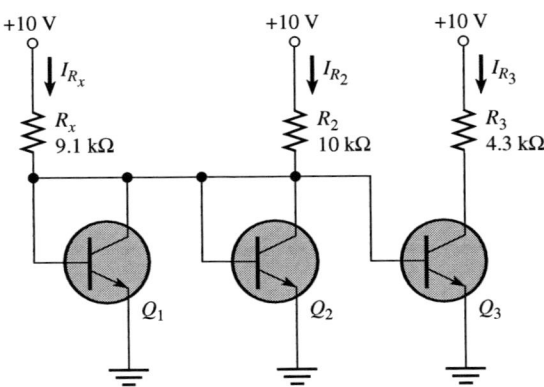

Figure 24-4

PROCEDURE

Part 1. JFET Current Source

a. Wire up the circuit of Fig. 24.1. Use $R_L = 51\ \Omega$. Measure and record the drain-source voltage.

V_{DS} (measured) = _____

b. Using the voltage measured in Part 1(a), calculate the load current.

$$I_{R_L} = \frac{V_{DD} - V_{DS}}{R_L}$$

$I_{R_L} =$ _____

c. Replace R_L with the resistors as listed in Table 24.1 and repeat Parts 1(a) and 1(b).

TABLE 24.1

R_L	20 Ω	51 Ω	82 Ω	100 Ω	150 Ω
V_{DS}					
I_{R_L}					

Exp. 24 / Procedure

Part 2. BJT Current Source

a. Calculate the current I_{R_L} through the load in the circuit of Fig. 24.2.

I_{R_L} (calculated) = _____

b. Wire up the circuit of Fig. 24.2. Measure and record the following voltages.

V_E (measured) = _____
V_C (measured) = _____

c. Calculate the emitter current and the current through the load.

I_{R_E} = _____
I_{R_L} = _____

d. Replace R_L with the resistors listed in Table 24.2 and repeat Parts 2(**a**) through 2(**c**).

TABLE 24.2

R_L	3.6 kΩ	4.3 kΩ	5.1 kΩ	7.5 kΩ
V_E				
V_C				
I_{R_E}				
I_{R_L}				

Part 3. Current Mirror

a. Calculate the mirror current in the circuit of Fig. 24.3.

I_x (calculated) = _____

b. Wire up the circuit of Fig. 24.3 and measure:

V_{B_1}(measured) = _____
V_{C_2} (measured) = _____
I_x = _____
I_{R_L} = _____

c. Change R_L to 3.6 kΩ and repeat Parts 3(**a**) and 3(**b**).

I_x (calculated) = _____
V_{B_1} (measured) = _____
V_{C_2}(measured) = _____
I_x = _____
I_{R_L} = _____

Part 4. Multiple Current Mirrors

a. Calculate the mirror current in the circuit of Fig. 24.4.

I_{R_x} (calculated) = _____

b. Wire up the circuit of Fig. 24.4 and measure:

V_{B_1} (measured) = _____
V_{C_2} (measured) = _____
V_{C_3} (measured) = _____
I_{R_x} = _____
I_{R_2} = _____
I_{R_3} = _____

c. Change R_L to 3.6 kΩ and repeat Parts 4(a) and 4(b).

I_{R_x} (calculated) = _____
V_{B_1} (measured) = _____
V_{C_2} (measured) = _____
V_{C_3} (measured) = _____
I_{R_x} = _____
I_{R_2} = _____
I_{R_3} = _____

Part 5. Computer Exercise

PSpice Simulation 24-1

The current mirror circuit shown is that of Fig. 24.3. Its property is that it can provide the same current over a range of resistance of RL.

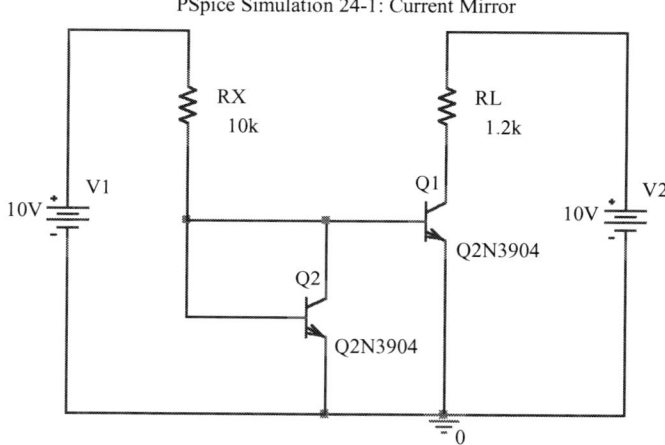

PSpice Simulation 24-1: Current Mirror

1. Perform a bias point analysis with resistor RL set to 1.2 kΩ.

2. Record the current through both RX and RL.

3. Are they of approximately equal value?

4. Perform a bias point analysis with resistor RL set to 3.6 kΩ.

5. Record the current through both RX and RL.

6. Are they of approximately equal value?

7. Compare the currents for the two analyses.

8. Are they of approximately equal value?

9. How does your data compare with the currents calculated from Eq. 24.3?

10. From your data, does the circuit perform as a current mirror?

11. Does the voltage VCE across Q1 change with a change in RL?

12. Is this circuit in effect a current source?

EXPERIMENT 25

Frequency Response of Common-Emitter Amplifiers

OBJECTIVE

To calculate and measure the frequency response of common-emitter amplifier circuits.

EQUIPMENT REQUIRED

Instruments

Oscilloscope
DMM
Function generator
DC power supply

Components

Resistors

(2) 2.2-kΩ
(1) 3.9-kΩ
(1) 10-kΩ
(1) 39-kΩ

Transistors

(1) 2N3904 (or equivalent general purpose npn)

Capacitors

(1) 1-μF
(1) 10-μF
(1) 20-μF

Exp. 25 / Frequency Response of Common-Emitter Amplifiers

EQUIPMENT ISSUED

Item	Laboratory serial no.
DC power supply	
Function generator	
Oscilloscope	
DMM	

RÉSUMÉ OF THEORY

The analysis of the frequency response of an amplifier can be considered in three frequency ranges: the low-, mid-, and high-frequency regions. In the low-frequency region the capacitors used for DC isolation (AC coupling) and bypass operation affect the lower cutoff (lower 3-dB) frequency. In the mid-frequency range only resistive elements affect the gain, the gain remaining constant. In the high-frequency region of operation, stray wiring capacitances and device inter-terminal capacitances will determine the circuit's upper cutoff frequency.

Lower Cutoff (lower 3-dB) Frequency: Each capacitor used will result in a cutoff frequency. The lower cutoff frequency at the network is then the largest of these lower cutoff frequencies. For the network of Fig. 25.1 the lower cutoff frequencies are as follows.

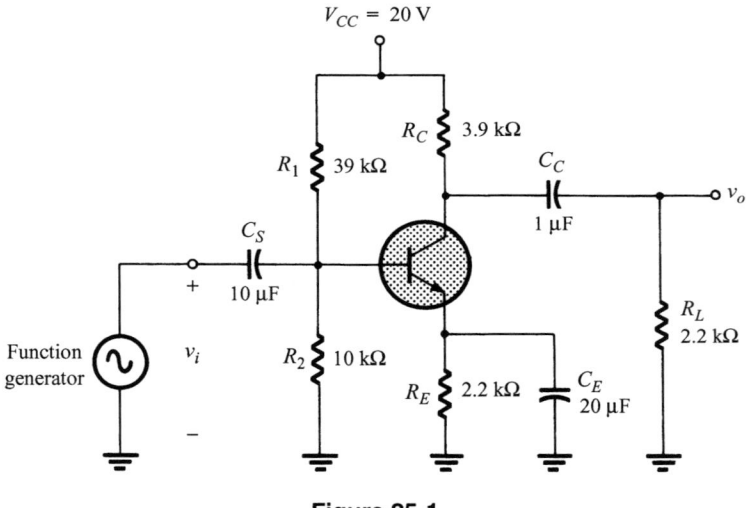

Figure 25-1

C_1: The cutoff frequency due to the input coupling capacitor is

$$f_{C_S} = \frac{1}{2\pi R_i C_i} \text{ Hz} \qquad \text{with: } R_i = R_1 ||R_2|| \beta r_e \qquad (25.1)$$

C_2: The cutoff frequency due to the output coupling capacitor is

$$f_{C_C} = \frac{1}{2\pi(R_C + R_L)C_i} \text{ Hz} \qquad (25.2)$$

C_E: The cutoff frequency due to the emitter bypass capacitor is

$$f_{C_E} = \frac{1}{2\pi R_e C_e} \text{Hz} \qquad \text{with } R_e = R_E \| r_e \qquad (25.3)$$

Upper Cutoff (upper 3-dB) Frequency: In the high-frequency range the amplifier gain is affected by the transistor's parasitic capacitances as follows:

At input connection of circuit:

$$f_{H_i} = \frac{1}{2\pi R_{Th_i} C_i} \text{ Hz} \qquad (25.4)$$

where

$$R_{Th_i} = R_1 \| R_2 \| \beta r_e$$

and C_i is

$$C_i = C_{w,i} + C_{be} + (1 + |A_v|)C_{bc}$$

$C_{w,i}$ = input wiring capacitance
A_v = voltage gain of amplifier at mid-band frequency
C_{be} = capacitance between transistor base-emitter terminals
C_{bc} = capacitance between transistor base-collector terminals

At output connection of circuit:

$$f_{H_o} = \frac{1}{2\pi R_{Th_o} C_o} \text{ Hz}$$

where

$$R_{Th_o} = R_C \| R_L$$

and

$$C_o = C_{w,o} + C_{ce}$$

$C_{w,o}$ = output wiring capacitance
C_{ce} = capacitance between transistor collector-emitter terminals

(We'll ignore the transistor's upper cutoff frequency, as it usually is greater than that due to wiring and inter-terminal capacitances.)

Keep in mind that the 3-dB cutoff frequencies are defined by 70.7% of the midband gain, or 0.707 $A_{v,\text{mid}}$. That is, once the midband gain is measured, the upper and lower cutoff frequencies are measured at the points at which the gain drops to 0.707, the midband gain at either upper or lower frequency.

PROCEDURE

Part 1. Low-Frequency Response Calculations

a. Using the specifications data for the transistor, record values:

C_{be} (specified) = _____
C_{bc} (specified) = _____
C_{ce} (specified) = _____

Enter values of typical wiring capacitance:

$C_{w,i}$ (approximated) = _____
$C_{w,o}$ (approximated) = _____

b. Using a characteristic curve tracer, beta measuring instrument, or value obtained from previous use in the lab, obtain the value of transistor beta.

β (measured) = _____

c. Calculate values of DC bias voltage and current for the circuit of Fig. 25.1.

V_B (calculated) = _____
V_E (calculated) = _____
V_C (calculated) = _____
I_E (calculated) = _____

Using the value of I_E, calculate the transistor dynamic resistance.

r_e (calculated) = _____

d. Calculate the magnitude of amplifier midband gain (under load) using

$$A_{v,\text{mid}} = \frac{R_C \| R_L}{r_e}$$

e. Calculate lower cutoff frequencies due to coupling capacitors and due to bypass capacity.

f_{C_S} (calculated) = _____
f_{C_C} (calculated) = _____
f_{C_E} (calculated) = _____

Part 2. Low-Frequency Response Measurements

a. Construct the network of Fig. 25.1. Record the actual resistor values in the space provided in Fig. 25.1, if desired. Adjust V_{CC} = 20 V. Apply an input AC signal, V_{sig} = 20 mV, at a peak frequency of f = 5 kHz. Observe the output voltage using a scope. If V_o shows any distortion, reduce V_{sig} until the output is undistorted.

b. Measure and record signals for undistorted operation.

V_{sig} (measured) = _____
V_o (measured) = _____

Calculate the circuit's mid-frequency voltage gain.

$A_{v,\text{mid}}$ = _____

Maintaining the input voltage at the level set above, vary the frequency and measure and record V_o to complete Table 25.1.

TABLE 25.1

f	50-Hz	100-Hz	200-Hz	400-Hz	600-Hz	800-Hz	1-kHz	2-kHz
V_o								

f	3-kHz	5-kHz	10-kHz
V_o			

Calculate the amplifier voltage gain for each frequency and complete Table 25.2.

TABLE 25.2

f	50-Hz	100-Hz	200-Hz	400-Hz	600-Hz	800-Hz	1-kHz	2-kHz
A_v								

f	3-kHz	5-kHz	10-kHz
A_v			

Part 3. High-Frequency Response Calculations

    a. Using the equations provided in the Résumé of Theory section calculate the upper cutoff frequencies and record below.

f_{H_i} (calculated) = _____

f_{H_o} (calculated) = _____

    b. Applying an input voltage which provides non-distorted output voltage, complete Table 25.3 measuring the resulting output voltage over a range of high frequency values.

V_i (measured) = _____

TABLE 25.3

f	10-kHz	50-kHz	100-kHz	300-kHz	500-kHz	600-kHz	700-kHz
V_o							

f	900-kHz	1-MHz	2-MHz
V_o			

Calculate the amplifier voltage gain (in dB units) and complete Table 25.4.

TABLE 25.4

f	10-kHz	50-kHz	100-kHz	300-kHz	500-kHz	600-kHz	700-kHz
A_v							

f	900-kHz	1-MHz	2-MHz
A_v			

Part 4. Gain versus Frequency

a. Using the semi-log paper of Fig. 25.2, plot the gain versus frequency over the full frequency range. Plot the actual points and connect to obtain the actual plot. Use straight-line approximation curves to obtain the Bode plot.

Figure 25-2

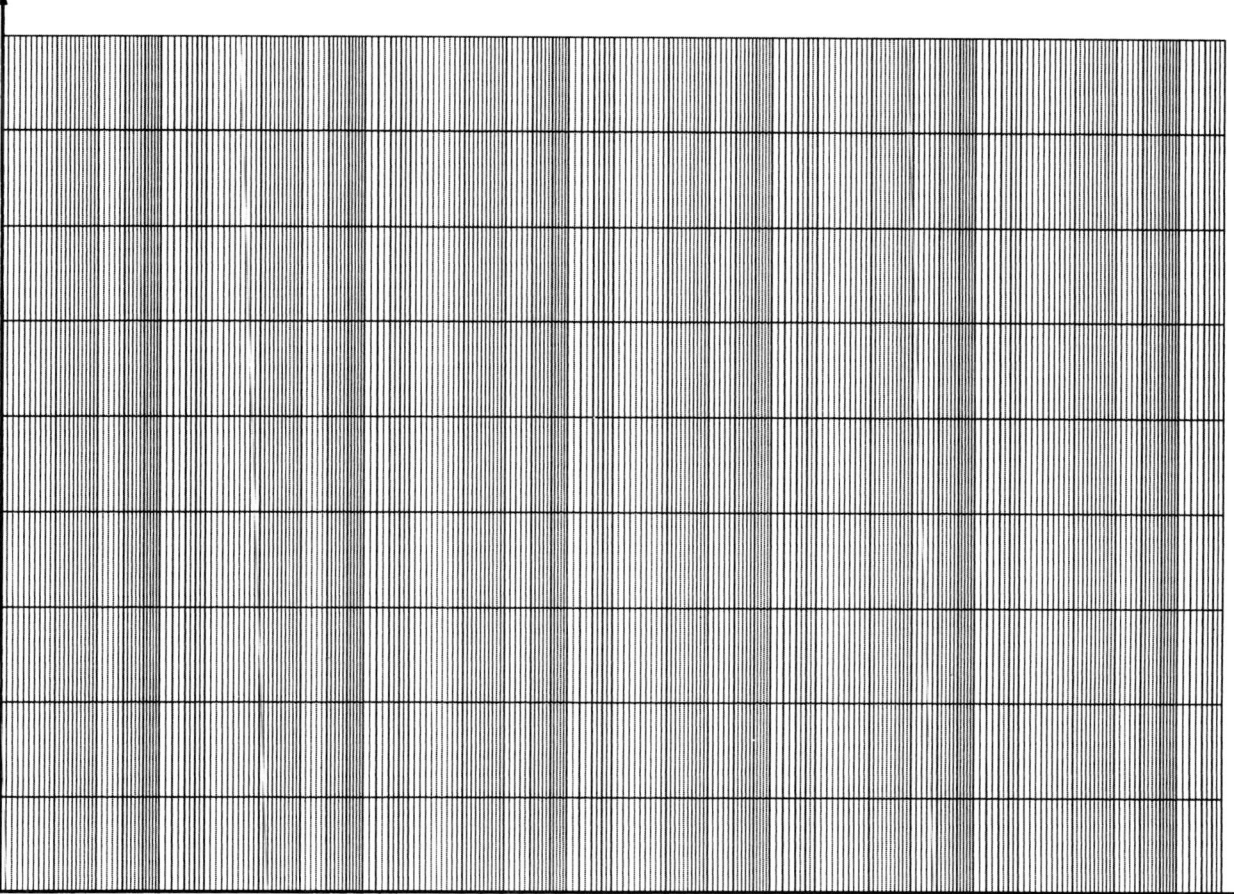

b. From the plot, obtain the lower and upper 3-dB frequency points and record below.

f_{-3dB} (measured) = _____

f_{+3dB} (measured) = _____

Compare the measured values with those calculated in Parts **1** and **3**.

Part 5. Computer Exercise

PSpice Simulation 25-1

The PSpice circuit used in this simulation is that of Fig. 25.1. We shall obtain the frequency response of this amplifier. Of particular interest are its loaded midband gain, the lower and upper corner frequencies, and the bandwidth. It is strongly recommended that this simulation be performed prior to the laboratory exercise. This will provide a benchmark against which to check the laboratory data obtained.

PSpice Simulation 25-1: Frequency Response of Common Emitter Amplifier

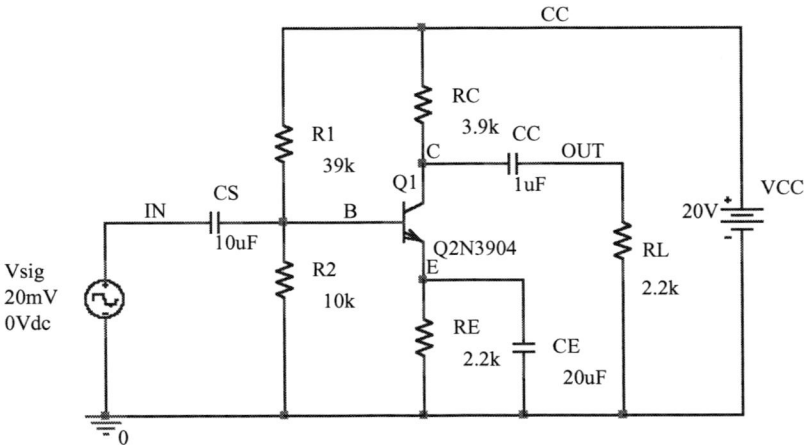

Vsig is an AC voltage source that permits an AC sweep (frequency) analysis of this circuit. Perform the stated analysis steps and answer the questions:

Exp. 25 / Part 5. Computer Exercise

1. Perform a bias point analysis.

2. From its data, compute the dynamic resistance and calculate the loaded midband gain of this amplifier.

3. Perform an AC sweep analysis. Set the AC Sweep Type to Logarithmic and select Decade. Set Start Frequency to 10 Hz and End Frequency to 1 GHz, and select 10 Points/Decade.

4. Obtain a plot of the numerical gain V(OUT)/V(IN).

5. Compare its midband gain to that calculated in step 2 above. Are they the same?

6. Obtain the bandwidth of this amplifier from the numerical gain using the two cursors in PSpice.

7. Obtain a plot of the logarithmic gain DB(V(OUT))/V(IN).

8. Is the logarithmic midband gain the equivalent of the numerical midband gain?

9. Compare the simulation data with the laboratory data. Explain any differences in the two data sets.

Class-A and Class-B Power Amplifiers

OBJECTIVE

To calculate and measure DC and AC voltages, and power input and output for both class-A and class-B power amplifiers.

EQUIPMENT REQUIRED

Instruments

Oscilloscope
DMM
Function generator
DC power supply

Components

Resistors

(1) 20-Ω
(1) 120-Ω, 0.5-W
(1) 180-Ω
(2) 1-kΩ, 0.5-W
(1) 10-kΩ

Capacitors

(3) 10-μF
(1) 100-μF

Transistors

(1) npn medium power, 15-W (2N4300 or equivalent)

(1) pnp medium power, 15-W (2N5333 or equivalent)
(2) Silicon diode

EQUIPMENT ISSUED

Item	Laboratory serial no.
DC power supply	
Function generator	
Oscilloscope	
DMM	

RÉSUMÉ OF THEORY

A class-A amplifier draws the same power from a voltage supply regardless of the signal applied. The input power is calculated from

$$P_i(DC) = V_{CC}I_{DC} = V_{CC}I_{CQ} \tag{26.1}$$

The signal power provided by the amplifier can be calculated using

$$P_o(AC) = \frac{V_C^2(\text{rms})}{R_C} = \frac{V_C^2(\text{peak})}{2R_C} = \frac{V_C^2(\text{p-p})}{8R_C} \tag{26.2}$$

with the amplifier's efficiency being

$$\%\eta = 100 \times \frac{P_o(AC)}{P_i(DC)}\% \tag{26.3}$$

A class-B amplifier draws no power if no input signal is applied. As the input signal increases, the amount of power drawn from the voltage supply and that delivered to the load both increase. The input power to a class-B amplifier is

$$P_i(DC) = V_{CC}I_{DC} = \frac{2V_{CC}V_C(p)}{\pi R_L} \tag{26.4}$$

The power provided by the amplifier can be calculated using:

$$P_o(AC) = \frac{V_L^2(\text{rms})}{R_L} = \frac{V_L^2(p)}{2R_L} = \frac{V_L^2(\text{p-p})}{8R_L} \tag{26.5}$$

The amplifier efficiency is calculated using Eq. 26.3.

Exp. 26 / Procedure

PROCEDURE

Part 1. Class-A Amplifier: DC Bias

a. Calculate the DC bias values for the circuit of Fig. 26.1.

$R_1 = $ _____
$R_2 = $ _____
$R_C = $ _____
$R_E = $ _____

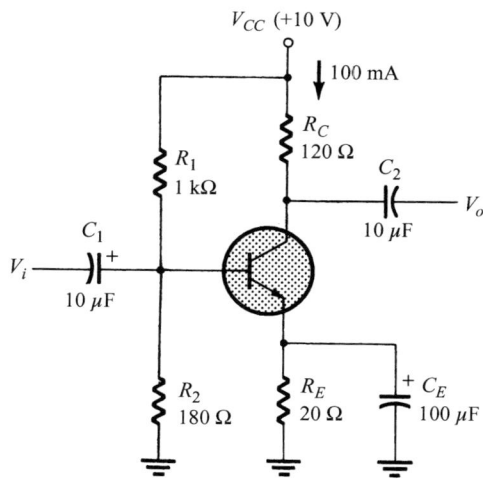

Figure 26-1

V_B (calculated) = _____
V_E (calculated) = _____
I_E (calculated) = I_C = _____
V_C (calculated) = _____

b. Construct the circuit of Fig. 26.1. If desired, measure and record actual resistor values in the space provided in Fig. 26.1. Adjust the supply voltage to $V_{CC} = 10$ V and measure and record DC bias voltages:

V_B (measured) = _____
V_E (measured) = _____
V_C (measured) = _____

Calculate the value of DC bias current:

$I_E = I_C = V_E/R_E =$ _____

Part 2. Class-A Amplifier: AC Operation

a. Using the DC bias values calculated in Part 1 and the equations given in the Résumé of Theory section, calculate power and efficiency values for the largest signal swing in the class-A amplifier of Fig. 26.1.

P_i (calculated) = _____

Using the largest signal swing around DC bias set in Part 1:

V_o (calculated) = _____
P_o (calculated) = _____
% η (calculated) = _____

b. Using the oscilloscope, adjust the input signal (f = 10 kHz) to obtain the largest undistorted output signal. Measure and record these input and output voltages.

V_i (measured) = _____
V_o (measured) = _____

c. Using the measured values, calculate the power and efficiency for the class-A amplifier of Fig. 26.1.

P_i = _____
P_o = _____
% η = _____

Compare the measured and calculated values of power and efficiency obtained in Parts 2(**b**) and 2(**c**).

Exp. 26 / Procedure

d. Reduce the input signal to one-half the level of Part 2(**b**). Measure and record the input and output voltages.

V_i (measured) = _____
V_o (measured) = _____

e. Calculate the input power, output power, and efficiency using half the input voltage used in Part 2(**a**).

P_i (calculated) = _____
P_o (calculated) = _____
% η (calculated) = _____

f. Using the measured values, calculate the power and efficiency for the class-A amplifier of Fig. 26.1.

P_i = _____
P_o = _____
% η = _____

Compare the measured and calculated values of power and efficiency obtained in Parts 2(**e**) and 2(**f**).

Part 3. Class-B Amplifier Operation

a. Calculate the power ratings for a class-B amplifier, as shown in Fig. 26.2, for $V_o = 1$ V, peak and $V_o = 2$ V, peak.

P_i (calculated) = _____
P_o (calculated) = _____
% η (calculated) = _____

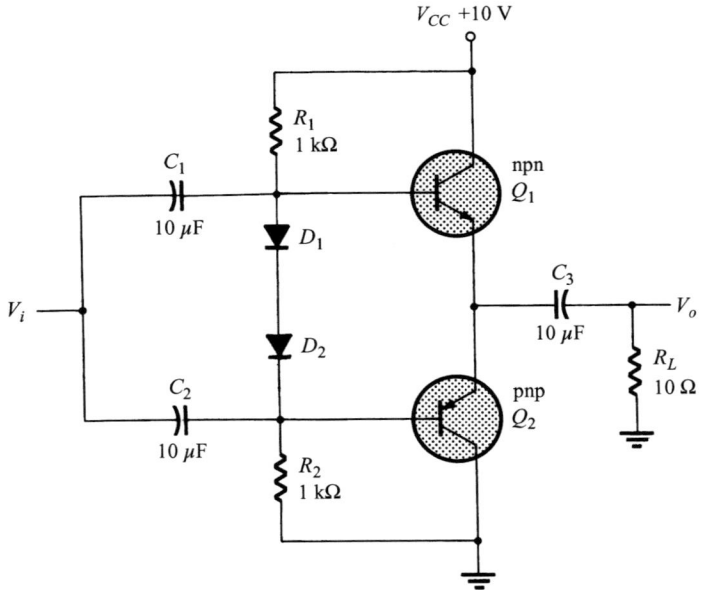

Figure 26-2

For $V_o = 2$ V, peak:

P_i (calculated) = _____
P_o (calculated) = _____
% η (calculated) = _____

b. Construct the circuit of Fig. 26.2. Adjust $V_{CC} = 10$ V. If desired, measure and record actual resistor values in the space provided in Fig. 26.2. Adjust the input until $V_o = 1$ V, peak. Measure and record AC voltages.

V_i (measured) = _____
V_o (measured) = _____

Exp. 26 / Procedure

Using the measured values, calculate input and output power, and circuit efficiency.

$P_i =$ _____
$P_o =$ _____
$\% \eta =$ _____

Compare values calculated in Part 3(**a**) with those measured in Part 3(**b**).

c. Adjust the input until $V_o = 2$ V, peak. Measure and record AC voltages.

V_i (measured) = _____
V_o (measured) = _____

Measure the average (DC) supply current from V_{CC}.

I_{DC} (measured) = _____

Using the measured values, calculate input and output power, and circuit efficiency:

$P_i =$ _____
$P_o =$ _____
$\% \eta =$ _____

Compare values calculated in Part 3(**a**) with those measured in Part 3(**c**).

Part 4. Computer Exercises

PSpice Simulation 26-1

The circuit shown is the class-A amplifier of Fig. 26.1. For this circuit, obtain the requested data and answer all stated questions.

PSpice Simulation 26-1: Class-A Amplifier

(Circuit diagram: Class-A amplifier with Vin source (VOFF = 0V, VAMPL = 20mV, FREQ = 10kHz) connected through C1 (10uF) to the base B of Q1 (Q2N3904). R1 = 1k from base to VCC, R2 = 180 from base to ground. RC = 120 from collector to VCC. RE = 20 from emitter to ground, bypassed by CE = 100uF. Output taken from collector through C2 (10uF) to OUT, with Rload = 10Meg to ground. VCC = 10V.)

1. Perform a DC bias point analysis.

2. From the obtained data, is the Q-point approximately in the center of the load line?

3. Is the voltage VCE about one-half the voltage VCC?

4. What is the DC power delivered by VCC to the circuit?

5. Which circuit element consumed the largest quiescent power?

6. Set the voltage of Vsig to 20 millivolts peak-to-peak at a frequency of 10 kHz.

7. Perform a Time Domain (transient) analysis of 200 microseconds duration.

8. From this data, is there an output voltage VOUT for 360° of the input voltage VIN?

9. Is the output voltage distorted?

10. What is the phase relationship between VIN and VOUT?

11. Compute the power provided by the amplifier from Eq. 26.2.

12. From this calculation, and the DC power supplied by VCC obtained above, calculate the efficiency of this amplifier at its operating point.

13. Reduce the voltage Vsig to 10 millivolts and repeat the analysis. In particular, what data remains the same, and what differs?

PSpice Simulation 26-2

The circuit shown is the class-B amplifier of Fig. 26.2.

PSpice Simulation 26-2: Class B Amplifier

For this simulation start with a bias point simulation. From this data, answer the following questions:

1. What is the DC input power provided by the voltage source VCC?

2. What circuit elements consume the largest amount of that power?

3. Are the two transistors biased at or near their cutoff points? If not, why not?

4. Is the DC voltage at node E2 about one-half the voltage of VCC?

5. What is the base-to-emitter voltage for both transistors?

6. What is the purpose of the two diodes in the circuit?

7. What is the voltage drop across each diode?

8. Apply a Vsig of 8 $V_{p\text{-}p}$ at a frequency of 1 kHz.

9. Perform a Time Domain (transient) analysis of 4 milliseconds duration.

10. What is the peak-to-peak output voltage VOUT?

11. Compute the efficiency of this circuit at its operating point.

12. Reduce Vsig to 4 V_{p-p} at the same frequency.

13. Repeat the Time Domain (transient) analysis.

14. Compute the present efficiency of the circuit and compare it to its previous value.

15. How does the simulation data compare to the experimental data?

Name _____
Date _____
Instructor _____

Differential Amplifier Circuits

OBJECTIVES

1. To calculate and measure DC and AC voltages in differential amplifier circuits.
2. To calculate the differential and common-mode gains of these amplifiers.

EQUIPMENT REQUIRED

Instruments

Oscilloscope
DMM
Function generator
DC power supply

Components

Resistors

(1) 4.3-kΩ
(4) 10-kΩ
(2) 20-kΩ

Transistors

(3) 2N3823, or equivalent

EQUIPMENT ISSUED

Item	Laboratory serial no.
DC power supply	
Function generator	
Oscilloscope	
DMM	

RÉSUMÉ OF THEORY

BJT Differential Amplifier

A differential amplifier is a circuit with plus (+) or minus (−) inputs. In typical operation, inputs that are opposite in-phase are amplified greatly, while inputs that are in-phase are canceled at the output. Figure 27.1 is the circuit of a simple BJT differential amplifier with plus (V_i^+) input and minus (V_i^-) input, and opposite outputs, V_{o1} and V_{o2}. Typically no capacitor is needed, the input signals being DC coupled, and the positive (V_{CC}) and negative (V_{EE}) supplies providing DC bias. Using the value of r_e assumed in this experiment to be the same for both transistors, the differential voltage gain is of magnitude

$$A_v = \frac{R_C}{2r_e} \tag{27.1}$$

The gain for signals which are common at both inputs is of magnitude

$$A_v = \frac{R_C}{2R_E} \tag{27.2}$$

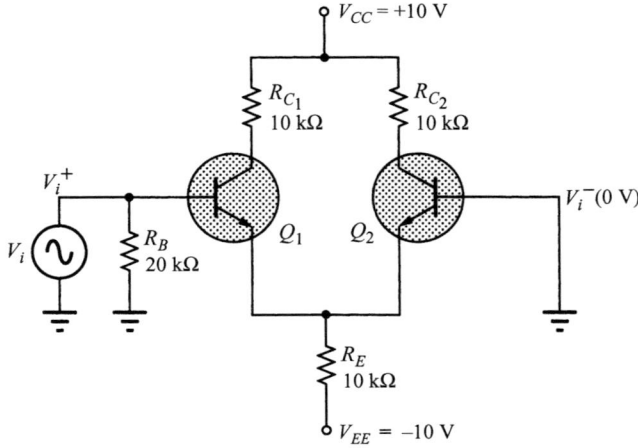

Figure 27-1

FET Differential Amplifier

For an FET differential amplifier the magnitude of the differential voltage gain can be calculated as

$$A_v = \frac{g_m R_D}{2} \tag{27.3}$$

PROCEDURE

Part 1. DC Bias of BJT Differential Amplifier

a. For the circuit of Fig. 27.1 calculate DC bias voltages and currents for one transistor.

V_B (calculated) = _____
V_E (calculated) = _____
V_C (calculated) = _____
I_E (calculated) = _____
r_e (calculated) = _____

b. Construct the circuit of Fig. 27.1. (Record measured value for all resistors in Fig. 27.1.) Set both supplies, V_{CC} = 10 V and V_{EE} = 10 V. Measure and record DC bias voltages for each transistor.

	Q_1		Q_2
V_B (measured) =	_____	V_B =	_____
V_E (measured) =	_____	V_E =	_____
V_C (measured) =	_____	V_C =	_____

Using measured values, determine

I_E = _____ I_E = _____
r_e = _____ r_e = _____

Compare values for each transistor to determine if they are well matched. Compare the values calculated in Part 1(a) with those measured in Part 1(b).

Part 2. AC Operation of BJT Differential Amplifier

a. Using Eqs. 27.1 and 27.2, calculate the differential and common-mode gain of the circuit in Fig. 27.1.

A_{v_d} (calculated) = _____
A_{v_c} (calculated) = _____

b. Apply input of V_i = 20 mV, rms at frequency f = 10 kHz to the plus (+) input and 0 V to the minus (−) input in the circuit of Fig. 27.1. Using a DMM, measure and record output voltages.

V_{o_1} (measured) = _____
V_{o_2} (measured) = _____

Calculate an average value of V_{o_d}.

$$V_{o_d} = \frac{V_{o_1} + V_{o_2}}{2}$$

V_{o_d} = _____

Calculate differential voltage gain.

$$A_{v_d} = \frac{V_{o_d}}{V_i}$$

A_{v_d} (measured) = _____

c. Apply common inputs of $V_i = 1$ V at a frequency of 10 kHz peak to both input terminals in the circuit of Fig. 27.1. Measure and record the output from one side of the circuit.

V_{o_c} (measured) = _____

Calculate the common voltage gain.

$$A_{v_c} = \frac{V_{o_c}}{V_i}$$

A_{v_c} (measured) = _____

Compare the voltage gains calculated in Part 2(a) with those measured in Parts 2(b) and 2(c).

Part 3. DC Bias of BJT Differential Amplifier with Current Source

a. Calculate DC bias voltages and currents for the amplifier of Fig. 27.2.

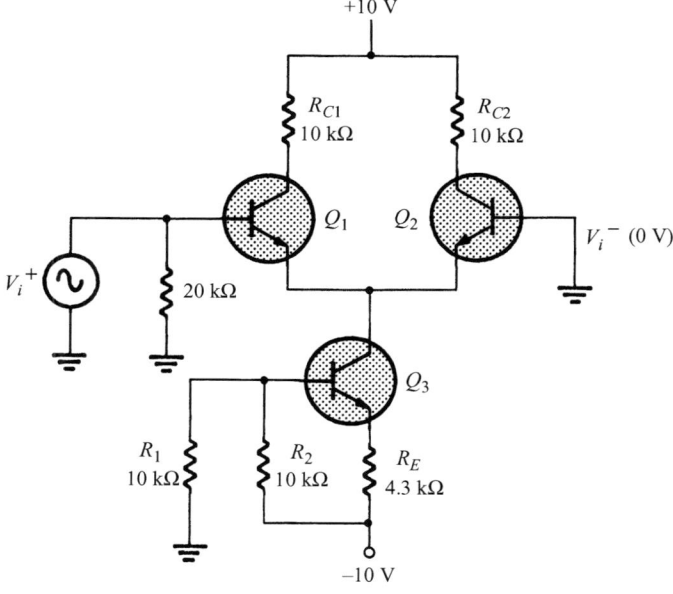

Figure 27-2

For either Q_1 or Q_2:

V_B (calculated) = _____
V_E (calculated) = _____
V_C (calculated) = _____
I_E (calculated) = _____
r_e (calculated) = _____

For Q_3:

V_B (calculated) = _____
V_E (calculated) = _____
V_C (calculated) = _____
I_E (calculated) = _____
r_e (calculated) = _____

b. With DC power turned off, construct the circuit of Fig. 27.2 (or just modify the circuit of Part 2). (Record measured resistor values in Fig. 27.2.) Restore DC power (10 V and –10 V) and measure DC bias voltages.

For transistors Q_1 and Q_2:

	Q_1	Q_2
V_B (measured) =	_____	V_B = _____
V_E (measured) =	_____	V_E = _____
V_C (measured) =	_____	V_C = _____

Using measured values, determine

I_E = _____ I_E = _____
r_e = _____ r_e = _____

Exp. 27 / Procedure

Compare values for both transistors to determine if the transistors are well matched.

For transistor Q_3:

V_B (measured) = _____
V_E (measured) = _____
V_C (measured) = _____

Using measured values, determine

$I_E =$ _____
$r_e =$ _____

Compare the values calculated in Part 3(**a**) with those measured in Part 3(**b**).

Part 4. AC Operation of Differential Amplifier with Transistor Current Source

a. Using Eq. 27.1, calculate

A_{v_d} (calculated) = _____

b. Apply input of $V_i^+ = 10$ mV, rms at frequency $f = 10$ kHz. Measure and record AC voltages.

V_{o_d} (measured) = _____

$$A_{v_d} = \frac{V_{o_d}}{V_i}$$

A_{v_d} (measured) = _____

c. Apply common input of $V_i = 1$ V, rms at frequency $f = 10$ kHz to both input terminals in the circuit of Fig. 27.2. Measure and record the output from one side of the circuit.

V_{o_c} (measured) = _____

Calculate the common voltage gain.

$$A_{v_c} = \frac{V_{o_c}}{V_i}$$

A_{v_c} (measured) = _____

d. Using the AC coupled input of the oscilloscope, measure and record the waveforms at each output and at the common-emitter point of the circuit. Record the waveforms in Fig. 27.3 showing proper phase relations.

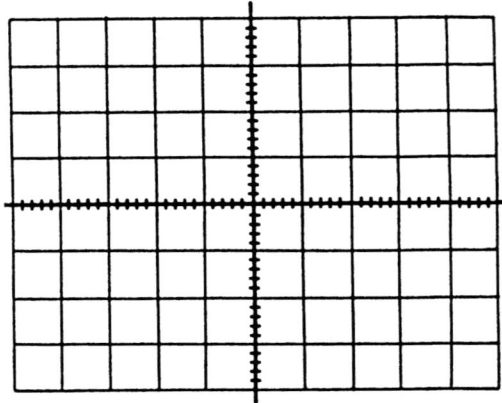

Vertical sensitivity = _____

Horizontal sensitivity = _____

Figure 27-3

Part 5. JFET Differential Amplifier

a. Obtain the values of I_{DSS} and V_P for each of the transistors in the circuit of Fig. 27.4 using previous procedures in Experiments 12 or 13. Record the values obtained.

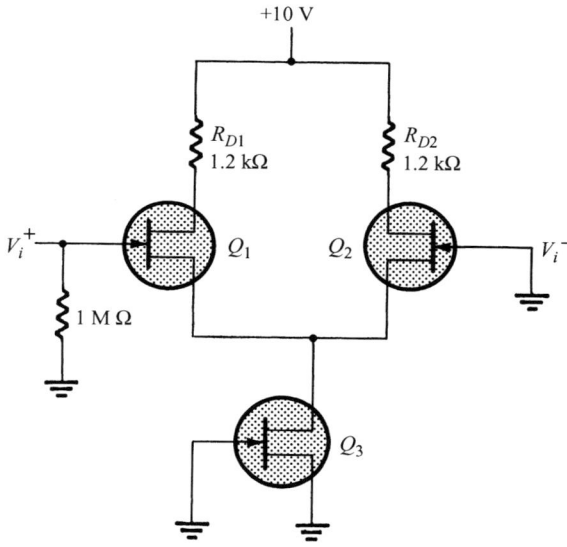

Figure 27-4

For Q_1:

I_{DSS} = _____
V_P = _____

For Q_2:

I_{DSS} = _____
V_P = _____

For Q_3:

I_{DSS} = _____
V_P = _____

b. Calculate the DC bias voltages and currents for the circuit of Fig. 27.4 using the values obtained in Part 5(a).

V_{D_1} (calculated) = _____
V_{D_2} (calculated) = _____
V_{S_1} (calculated) = _____

c. Construct the circuit of Fig. 27.4. (Record measured resistor values in Fig. 27.4.) Measure and record the DC voltages.

V_{G_1} (measured) = _____
V_{D_1} (measured) = _____
V_{D_2} (measured) = _____
V_{D_3} (measured) = _____

d. Calculate the value of the circuit differential voltage gain.

A_{v_d} (calculated) = _____

e. Using AC coupling, apply an input, $V_i^+ = 50$ mV, rms at frequency $f = 10$ kHz. Using a DMM measure, record the output voltages.

V_{o_1} (measured) = _____
V_{o_2} (measured) = _____

Determine AC differential voltage gain (using V_{o_1} and then V_{o_2}).

$A_{v_1}(d) = $ _____
$A_{v_2}(d) = $ _____

Compare the values of differential voltage gain measured in Part 5(**e**) with that calculated in Part 5(**d**).

f. For input of $V_i^+ = 50$ mV, peak, observe waveforms at all three transistors' drain terminals and record in Fig. 27.5 showing the proper phase relations.

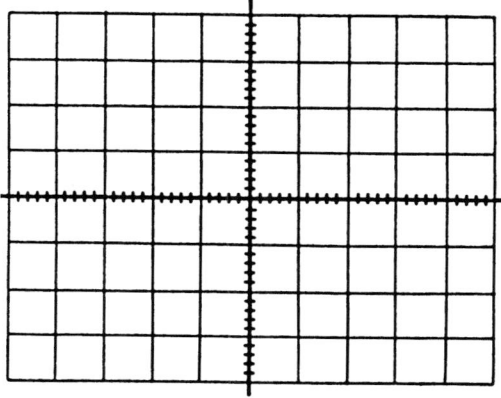

Vertical sensitivity = _____

Horizontal sensitivity = _____

Figure 27-5

$A_{v_c} =$ _____

Part 6. Computer Exercises

PSpice Simulation 27-1

The BJT differential amplifier shown is that of Fig. 27.1.

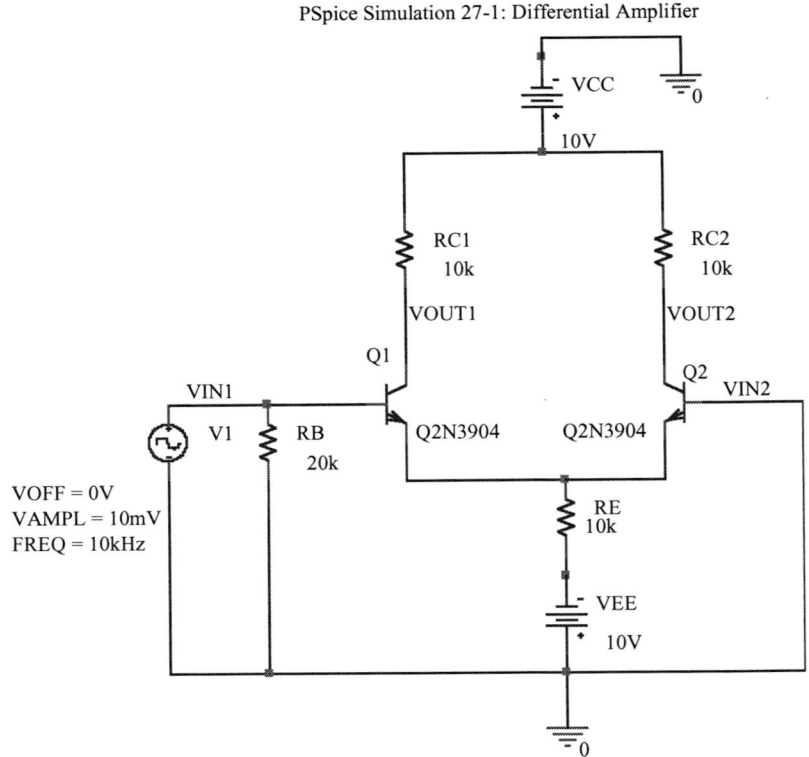

Start with a DC bias simulation to obtain the following data and answer the following questions:

1. What is the total DC power delivered to the circuit by the sources VCC and VEE?

2. Do both sources contribute equally to that power?

3. Obtain the collector and emitter voltages for Q1 and Q2.

4. Are they the same?

5. Obtain the collector currents for Q1 and Q2.

6. Are they the same?

7. Set the voltage of signal source V1 to 20 millivolts at a frequency of 10 kHz.

8. Perform a Time Domain (transient) analysis for 200 microseconds.

9. Obtain the Probe plots of V(OUT1) and V(OUT2).

10. What is their magnitude and relative phase shift?

11. What is the gain for the amplifier in the single-ended mode? Use the ratio of RMS(V(VOUT1))/RMS(V(VIN1)).
    Obtain a Probe plot of VOUT. Calculate the voltage gain from data. Note: VOUT = VOUT1 −VOUT2.

PSpice Simulation 27-1: Differential Amplifier

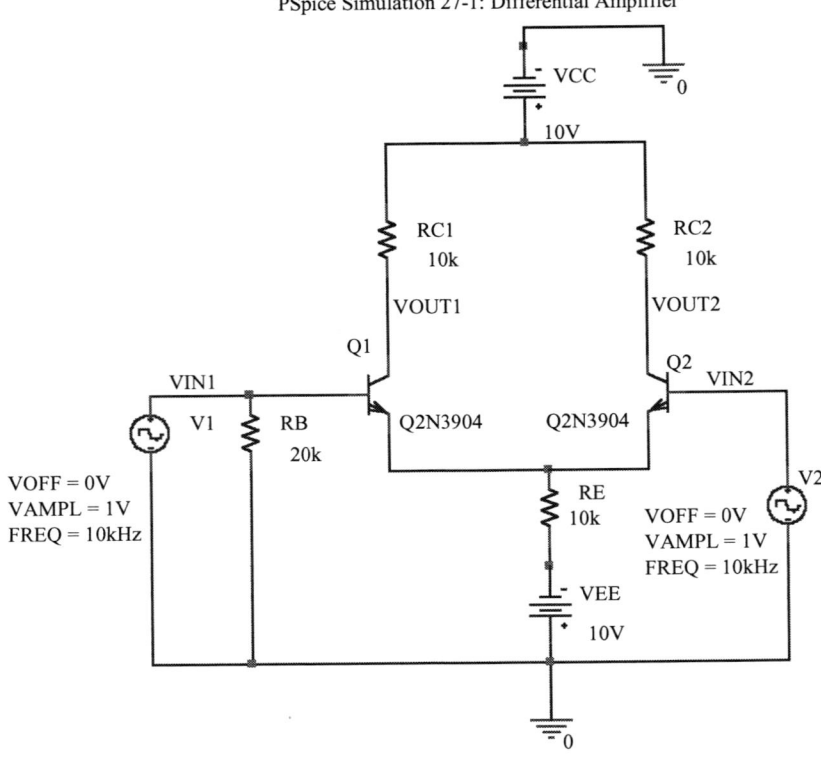

12. Modify the circuit as shown above for common-mode operation by adding signal source V2 and setting both signal sources to 1 V at 10 kHz.

13. Repeat the Time Domain (transient) analysis.

14. Obtain the Probe plots of V(OUT1) and V(OUT2).

15. What is their magnitude and relative phase shift?

16. What is the gain for the amplifier in the common mode? Use the same voltage ratio as in step 11.

PSpice Simulation 27-2

The BJT differential amplifier with current source shown is that of Fig. 27.2.

PSpice Simulation 27-2: Differential Amplifier with Current Source

Begin with a DC bias simulation to obtain the requested data and answer the following questions:

1. What is the total DC power delivered by the voltage sources VCC and VEE?

2. What are the collector voltages for Q1 and Q2? Are they the same?

3. Obtain the collector currents for Q1, Q2, and Q3.

4. What are their relative magnitudes?

5. Set the AC voltage of V1 to 10 millivolts at 10 kHz.

6. Perform a time (transient) analysis for 200 microseconds.

7. Obtain the Probe plots of V(OUT1) and V(OUT2).

8. What is their magnitude and relative phase shift?

9. Determine the gain of the amplifier in the single-ended mode. Use the same ratio as in PSpice simulation 27-1, step 11.

10. Modify the circuit as shown for common-mode operation.

Exp. 27 / Part 6. Computer Exercises

PSpice Simulation 27-2: Differential Amplifier with Current Source

11. Repeat the Time Domain (transient) analysis.

12. Obtain the Probe plots of V(OUT1) and V(OUT2). To obtain their smooth plots, specify a Maximum print step of 0.01 us.

13. What is their magnitude and relative phase shift?

14. What is the gain for the amplifier in the common mode? Use the same voltage ratio as in step 9 above.

Name _____
Date _____
Instructor _____

EXPERIMENT 28

Op-Amp Characteristics

OP-AMP SLEW RATE AND COMMON MODE REJECTION RATIO

OBJECTIVES

1. To measure the Slew Rate of an µ741 operational amplifier and to calculate its common mode rejection ratio (CMR).
2. To run a PSpice analysis to obtain the slew rate of the µ741 operational amplifier and its CMR.
3. To compare the results of these analysis with the experimental results.
4. To compare our data with their published values.

EQUIPMENT REQUIRED

Instruments

Oscilloscope (dual trace)
Signal generator
DC power supply
DMM

Components

Resistors

(2) 100-Ω
(2) 10-kΩ
(2) 100-kΩ

IC

(1) µ741 (or equivalent) op-amp

EQUIPMENT ISSUED

Item	Laboratory serial no.
Oscilloscope(dual trace)	
Signal generator	
DC power supply	
DMM	

RÉSUMÉ OF THEORY

Determining the Slew Rate

The maximum rate of change that the output voltage undergoes in response to a step input voltage is defined as the slew rate of the op-amp. Slew rates are defined in units of volts/microseconds, V/µs. The slew rate is dependent upon the frequency response of the internal stages of the op-amp. The larger its value, the broader is the frequency response of the op-amp.

It is calculated as follows: SR = $\Delta V_{out}/\Delta t$ = _____ V/µs

Where $\Delta V_{out} = +V_{max} - (-V_{max})$. Δt is defined as the time interval between the positive and negative extremes of V_{out}.

Determining the Common Mode Rejection Ratio

When different input voltages are applied to the input terminals of an op-amp, and incidentally one of them could be zero volts, the op-amp operates in the differential mode. An amplified voltage will appear at its output terminal. If the same signal is applied to both input terminals of the op-amp, it operates in the common mode. Under ideal condition, its output voltage should be zero. However, the non-ideal nature of physical devices will result in the presence of a small output voltage under common mode operation. The ratio of the differential voltage gain, A(dif), and the common mode gain, A(cm), is defined as the common mode rejection ratio, CMR. That ratio is expressed in dB.

It is calculated as follows: $\text{CMR (dB)} = 20\log\left[\dfrac{A(dif)}{A(cm)}\right]$

In this formulation: $A(\text{dif}) = R_1/R_2$ (see Fig. 28-2)
$A(\text{cm}) = V(\text{out})/V(\text{in})$ (see Fig. 28-2)

Exp. 28 / Procedure

PROCEDURE

Part 1. Determining the Slew Rate

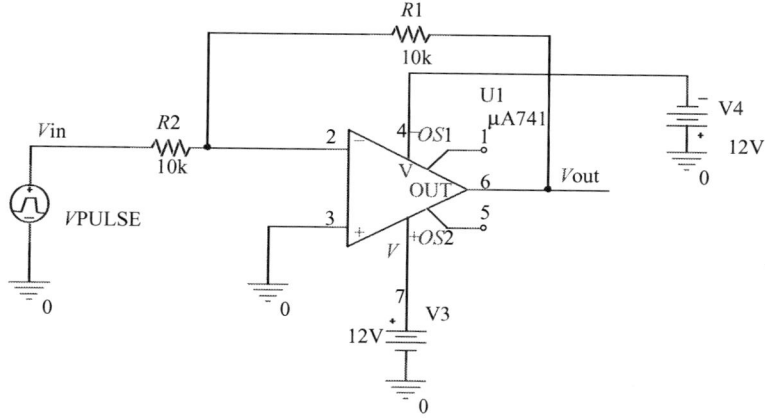

Figure 28-1

a. Wire the circuit shown in Fig. 28-1.

b. Connect terminals 4 and 7 of the op-amp to the −12 V and the +12 V terminals, respectively, of the DC power supply.

c. Connect the oscilloscope channel 1 to Vin. Set its vertical sensitivity to 2 V/division. Connect channel 2 to Vout. Set its vertical sensitivity to 1V/division. Select a horizontal sensitivity (time base) of 10 us/division. Use AC coupling.

d. Set the square wave input voltage of VPULSE to 5 Vp-p and its frequency to 10 kHz. Apply power.

e. Observe the traces of Vin and Vout on the scope. *Hint:* Vin will be a square wave with 0 rise and fall times. Vout will be a trapezoidal wave with finite rise and fall times.

f. Measure the peak-peak voltage of Vout. Define this as ΔV.

g. Measure the time interval it takes to go from one extreme of V_{out} to the other. Define this as Δt.

h. Calculate and record the slew rate:

$$SR = \Delta V/\Delta t \text{ volt/microseconds} = \underline{\hspace{2cm}} \text{ V/µs}.$$

i. Compare this value with that published for the µ741 op-amp.

Part 2. Determining the Common Mode Rejection Ratio

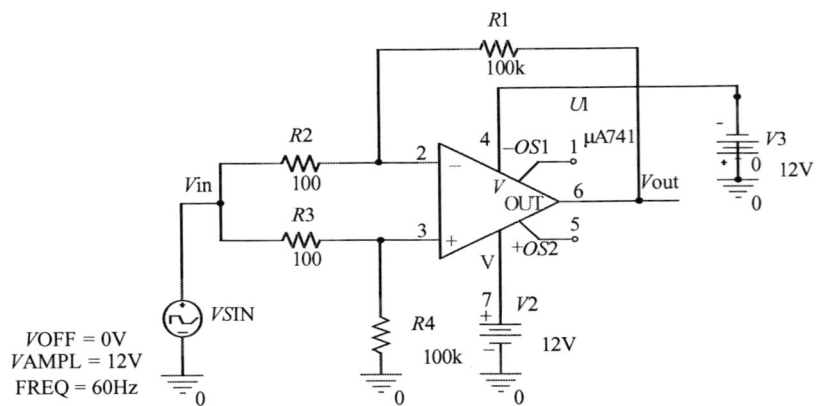

Figure 28-2

a. Wire the circuit shown in Fig. 28-2.

b. Connect terminals 4 and 7 of the op-amp, respectively, to the −12 V and +12 V terminals of the power supply.

c. Connect the oscilloscope channel 1 to V_{in}. Set its vertical sensitivity to 2 V/division. Connect channel 2 to V_{out}. Set its vertical sensitivity to 0.02 V/division.

Exp. 28 / Procedure

    d. Use AC coupling for both channels.

    e. Set the time base to 5 ms/division.

    f. Set the voltage of VSin to 12 Vp-p at a frequency of 60 Hz. Apply power.

    g. Use the DMM to measure the RMS voltages of Vin and Vout.

    h. Calculate the common mode voltage gain, $A(cm)$.

$$A(cm) = \frac{V(out)}{V(in)} = \underline{}$$

    i. Calculate the differential voltage gain, $A(dif)$

$$A(dif) = \frac{R1}{R2} = \underline{}$$

    j. Calculate the common mode rejection ratio

$$CMR(dB) = 20 \log \left[\frac{A(dif)}{A(cm)} \right] = \underline{} \text{ dB}$$

    k. Compare this value with that published for the μ741 op-amp.

Part 3. Computer Exercises

PSpice Simulation: Determining the Slew Rate

a. Modify the circuit of Fig. 28-1 as shown. Terminals 4 and 7 of the op-amp remain at their respective saturation voltages as in Fig. 28-1. The PSpice voltage source VPULSE has been connected as shown. Its parameters are set as indicated. Select a Time Domain (Transient) analysis of 0.2 ms duration. Run the analysis.

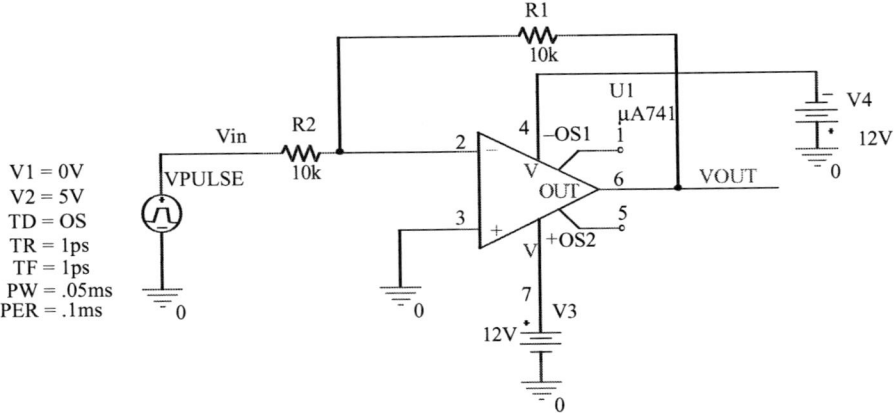

Figure 28-3

b. Set the time axis of the Probe plot to 200 µs. Obtain the trace of V(Vout) on a Probe plot. Place the cursors A1 and A2, respectively, at the maximum and the minimum voltage of V(Vout).

$V(Vout)_{max}$ = _____ volts

$V(Vout)_{min}$ = _____ volts

c. Read the time interval between these voltages and record it:

Time interval Δt = _____ µs

d. Using the maximum and minimum voltage of V(Vout) and the time interval, compute the value of the slew rate (SR) from the ratio

$SR = [V(Vout)max - V(Vout)min]/\Delta t$ = _____ µs

e. Compare this value with both the experimental and published values for the slew rate of the µ741 op-amp.

PSpice Simulation: Determining the Common Mode

a. Modify the circuit of Fig. 28-2 as shown. Terminals 4 and 7 of the op-amp remain at their respective saturation voltages as in Fig. 28-2. The signal generator of the laboratory experiment has been modeled by a VSIN voltage source. Its parameters are set as shown. Select a Time Domain (Transient) analysis of 34 ms duration. Run the analysis.

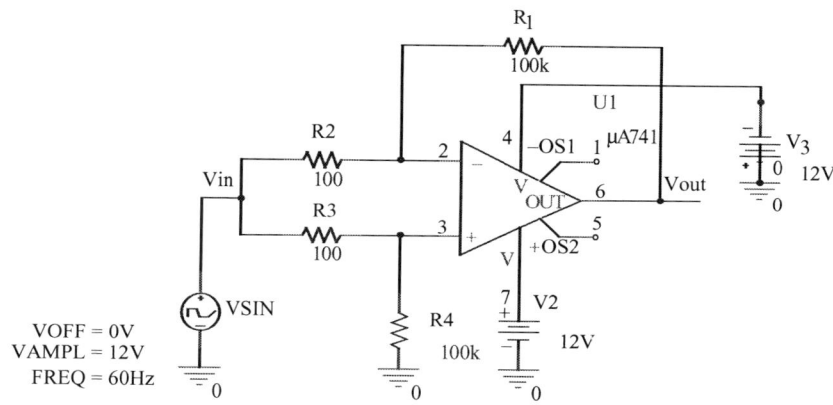

Figure 28-4

b. Obtain the traces of the RMS voltages of V(Vin) and V(Vout) on a Probe plot. Place cursor A1 at $t = 20$ ms and measure their respective amplitudes. From that, compute their common-mode voltage gain, $A(cm)$:

$$A(cm) = \frac{V_{(out)}}{V_{(in)}} = \underline{}$$

c. Compute the differential voltage gain *A(dif)*:

$$A(dif) = \frac{R1}{R2} = \underline{\hspace{2cm}}$$

d. Compute the common mode rejection ratio, CMR(dB):

$$\text{CMR(dB)} = \left[\frac{A(dif)}{A(cm)}\right] = \underline{\hspace{2cm}}$$

e. Compare this value with both the experimental and published values for the µ741 op-amp.

Name _____
Date _____
Instructor _____

EXPERIMENT 29

Linear Op-Amp Circuits

OBJECTIVES

1. To measure DC and AC voltages in linear op-amp circuits.
2. To compute the voltage gains of various op-amps.

EQUIPMENT REQUIRED

Instruments

Oscilloscope
DMM
Function generator
DC power supply

Components

Resistors

(1) 20-kΩ
(3) 100-kΩ

ICs

(1) 741 op-amp

EQUIPMENT ISSUED

Item	Laboratory serial no.
DC power supply	
Function generator	
Oscilloscope	
DMM	

RÉSUMÉ OF THEORY

The op-amp is a very high gain amplifier with inverting and noninverting inputs. It can be used to provide a much smaller but exact gain set by external resistors or to sum more than one input, each input causing a desired voltage gain.

As an inverting amplifier, the resistors are connected to the inverting input as shown in Fig. 29.1 with output voltage

$$V_o = -\frac{R_o}{R_i} V_i \qquad (29.1)$$

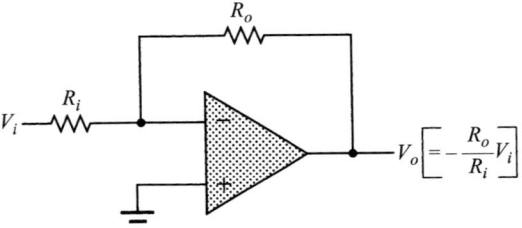

Figure 29-1

A noninverting amplifier is provided by the circuit of Fig. 29.2 with output voltage given by

$$V_o = \left(1 + \frac{R_o}{R_i}\right) V_i \qquad (29.2)$$

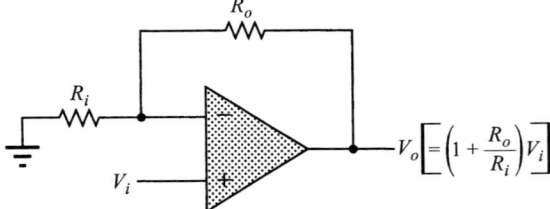

Figure 29-2

Connecting the output back to the inverting input as in Fig. 29.3 provides a gain of exactly unity:

$$V_o = V_i \qquad (29.3)$$

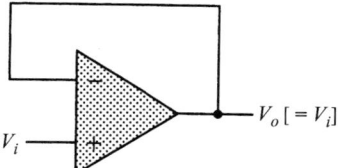

Figure 29-3

More than one input can be connected through separate resistors as shown in Fig. 29.4, with the output voltage then

$$V_o = -\left(\frac{R_o}{R_1}V_1 + \frac{R_o}{R_2}V_2\right) \qquad (29.4)$$

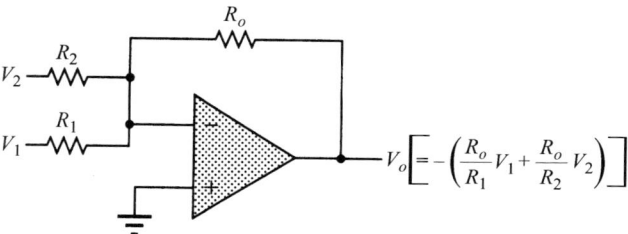

Figure 29-4

PROCEDURE

Part 1. Inverting Amplifier

a. Calculate the voltage gain for the amplifier circuit of Fig. 29.5.

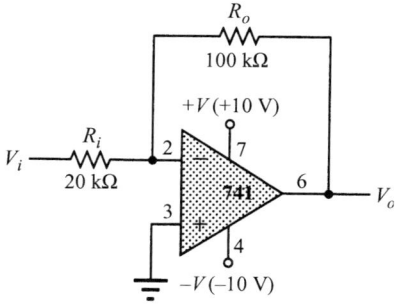

Figure 29-5

V_o/V_i (calculated) = _____

b. Construct the circuit of Fig. 29.5. (Measure and record resistor values in Fig. 29.5.) Apply an input of $V_i = 1$ V, rms ($f = 10$ kHz). Using a DMM, measure and record the output voltage.

V_o (measured) = _____

Calculate the voltage gain using measured values:

A_v = _____

Compare the gain calculated in Part 1(**a**) with that measured in Part 1(**b**).

c. Replace R_i with a 100-kΩ resistor. Calculate V_o/V_i.

V_o/V_i (calculated) = _____

For input of $V_i = 1$ V, rms measure and record V_o.

V_o (measured) = _____

Calculate A_v.

A_v = _____

Compare the calculated and measured values of the voltage gain.

d. Using the oscilloscope, observe and sketch the input and output waveforms in Fig. 29.6.

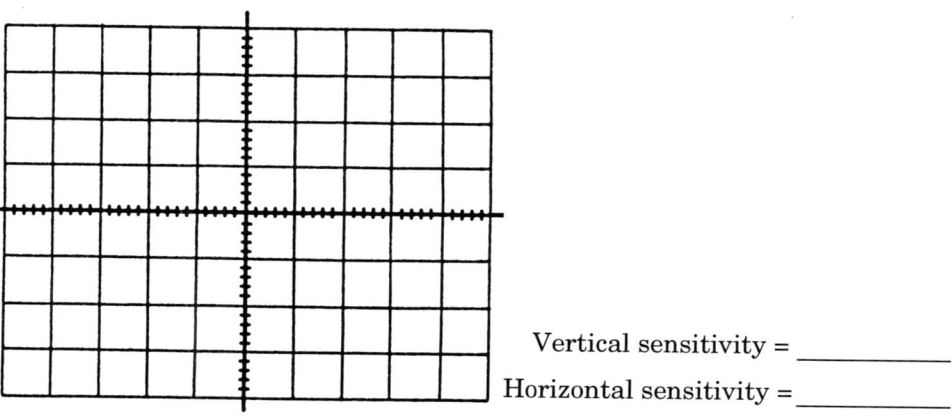

Vertical sensitivity = _____

Horizontal sensitivity = _____

Figure 29-6

Part 2. Noninverting Amplifier

a. Calculate the voltage gain of the noninverting amplifier in Fig. 29.7.

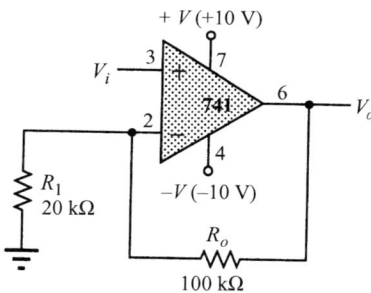

Figure 29-7

A_v (calculated) = _____

b. Construct the circuit of Fig. 29.7. Apply an input of $V_i = 1$ V, rms ($f = 10$ kHz). Using a DMM, measure and record the output voltage.

V_o (measured) = _____

Calculate the voltage gain of the circuit using measured voltages.

$$V_o/V_i = \underline{\hspace{2cm}}$$

Compare the voltage gain calculated in Part 2(**a**) with that measured in Part 2(**b**).

c. Replace R_i with a 100-kΩ resistor and repeat Parts 2(**a**) and 2(**b**).

$$A_v \text{ (calculated)} = \underline{\hspace{2cm}}$$
$$V_o \text{ (measured)} = \underline{\hspace{2cm}}$$
$$V_o/V_i = \underline{\hspace{2cm}}$$

Compare the calculated voltage gain with that measured.

d. Using the oscilloscope, observe and sketch the input and output waveforms in Fig. 29.8.

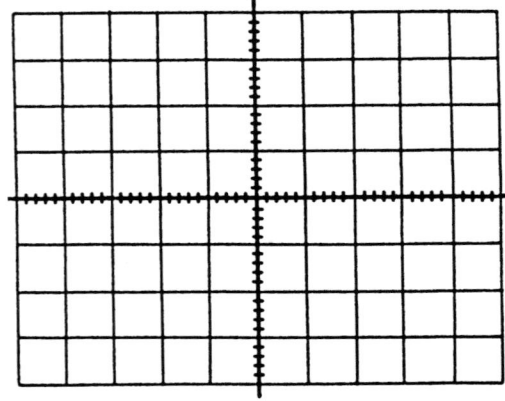

Vertical sensitivity = \underline{\hspace{2cm}}

Horizontal sensitivity = \underline{\hspace{2cm}}

Figure 29-8

Part 3. Unity-Gain Follower

a. Construct the circuit of Fig. 29.9. Apply an input signal of $V_i = 2$ V, rms ($f = 10$ kHz). Using a DMM, measure and record the input and output voltages.

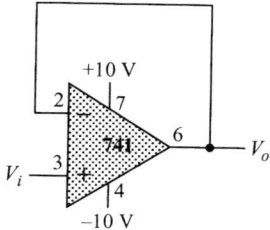

Figure 29-9

V_i (measured) = _____
V_o (measured) = _____

Compare the circuit voltage gain, V_o/V_i, with the theoretical unity gain.

Part 4. Summing Amplifier

a. Calculate the output voltage for the circuit of Fig. 29.10 (see Fig. 29.4) with inputs of $V_1 = V_2 = 1$ V, rms.

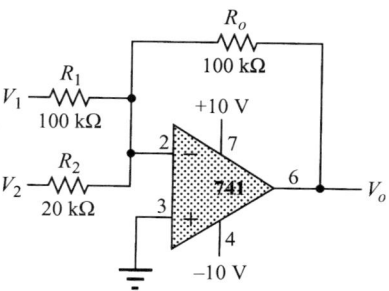

Figure 29-10

V_o (calculated) = _____

b. Construct the circuit of Fig. 29.10. Apply inputs of $V_1 = V_2 = 1$ V, rms ($f = 10$ kHz). Measure and record the output voltage.

V_o (measured) = _____

Compare the output voltage calculated in Part 4(**a**) and that measured in Part 4(**b**).

c. Change R_2 to 100 kΩ. Repeat Parts 4(**a**) and 4(**b**).

V_o (calculated) = _____
V_o (measured) = _____

Compare the calculated output voltage with that measured.

Part 5. Computer Exercises

PSpice Simulation 29-1

The inverting amplifier shown here is that of Fig. 29.5.

PSpice Simulation 29-1: Inverting Amplifier

Perform a Time Domain (transient) analysis of 200 microseconds duration. Obtain the requested data and answer all questions:

1. Obtain Probe plots of both the input voltage VIN and the output voltage VOUT.

2. What are their respective peak amplitudes?

3. Compute the theoretical gain of this amplifier using Eq. 29.1.

4. From the Probe plots, compute the gain of this amplifier as the ratio of the peak voltage of VOUT over the peak voltage of VIN.

5. Compare this result to the theoretical gain calculated above. Are they in agreement?

6. What is the relative phase of VOUT versus VIN?

7. Is the phase shift consistent with that of an inverter amplifier?

PSpice Simulation 29-2

The non-inverting amplifier is that of Fig. 29.7.

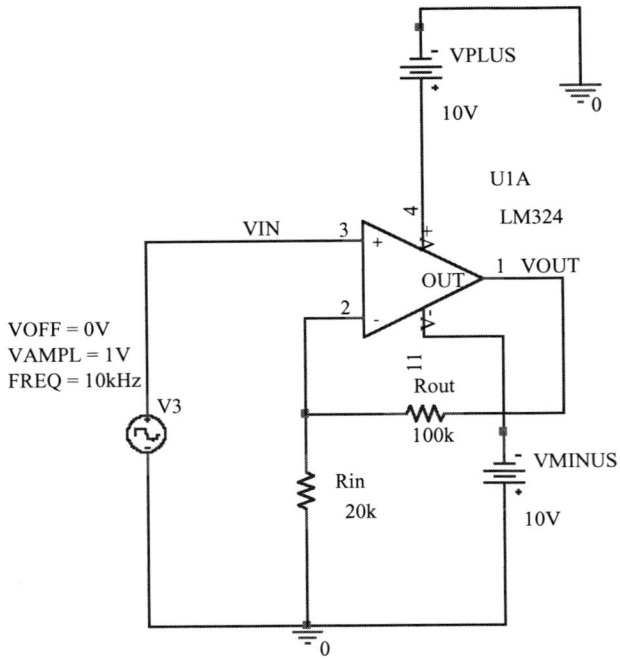

PSpice Simulation 29-2: Noninverting Amplifier

Perform a Time Domain (transient) analysis of 200 microseconds duration. Obtain the requested data and answer all questions:

1. Obtain Probe plots of both the input voltage VIN and the output voltage VOUT.

2. What are their respective peak amplitudes?

3. Compute the theoretical gain of this amplifier using Eq. 29.2.

4. From the Probe plots, compute the gain of this amplifier as the ratio of the peak voltage of VOUT over the peak voltage of VIN.

5. Compare this result to the theoretical gain calculated above. Are they in agreement?

6. What is the relative phase of VOUT versus VIN?

7. Is the phase shift consistent with that of a non-inverting amplifier?

EXPERIMENT 30

Active Filter Circuits

OBJECTIVE

To calculate and measure the critical frequencies and to measure AC voltages as a function of frequency of various types of active filter circuits.

EQUIPMENT REQUIRED

Instruments

Oscilloscope
DMM
Function generator
DC power supply

Components

Resistors

(5) 10-kΩ
(1) 100-kΩ

Capacitors

(2) 0.001-μF

Transistors and ICs

(1) 301 IC, or equivalent

EQUIPMENT ISSUED

Item	Laboratory serial no.
DC power supply	
Function generator	
Oscilloscope	
DMM	

RÉSUMÉ OF THEORY

Op-amps can be used to build active filter circuits for use as low-pass, high-pass, or band-pass filter operation. Filter operation provides the output of the filter dropoff as a function of frequency to 0.707 of the starting value at the cutoff frequency. This is a drop of 3 dB. The rate of amplitude decrease is at 6-dB per octave (half or twice frequency), which is the same as 20-dB per decade (tenfold larger or smaller frequency).

Low-Pass Filter

A low-pass active filter passes frequencies below the filter cutoff frequency. The circuit of Fig. 30.1 shows the connection of an op-amp unit as a low-pass filter, the low-cutoff frequency determined by

$$f_L = \frac{1}{2\pi R_1 C_1} \text{ Hz} \qquad (30.1)$$

The output V_O drops off at 6 dB/octave or 20 dB/decade above the cutoff frequency.

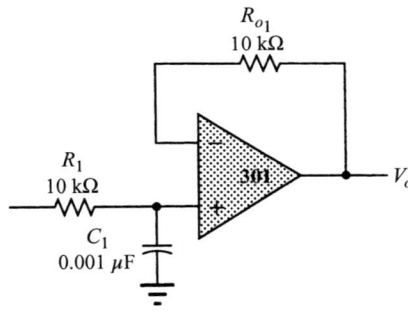

Figure 30-1

High-Pass Filter

A high-pass filter, as shown in Fig. 30.2, maintains the output amplitude at frequencies above a high-cutoff frequency determined by

$$f_H = \frac{1}{2\pi R_2 C_2} \text{ Hz} \qquad (30.2)$$

The output V_O drops off at 6 dB/octave or 20 dB/decade below the cutoff frequency.

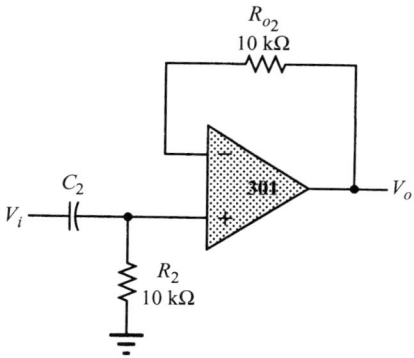

Figure 30-2

Band-Pass Filter

A band-pass filter circuit, as shown in Fig. 30.3, passes the input signal only for frequencies within a band of frequencies. The circuit shown is basically low-pass and high-pass active filters in series. The band-pass low- and high-cutoff frequencies are then calculated using Eqs. 30.1 and 30.2.

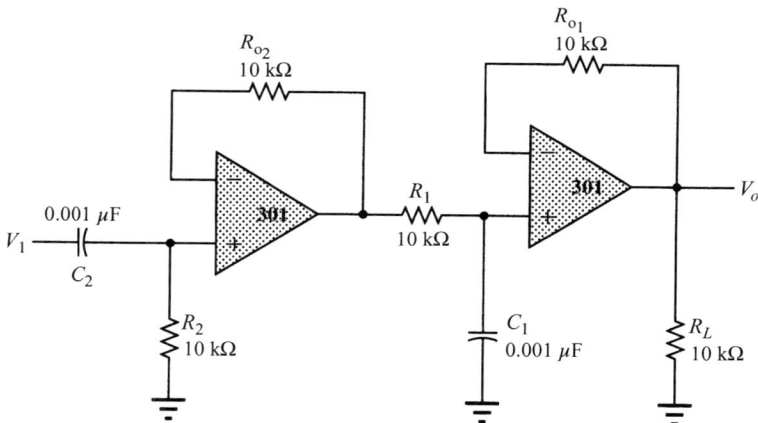

Figure 30-3

PROCEDURE

Part 1. Low-Pass Active Filter

a. For the circuit of Fig. 30.1 calculate the low-cutoff frequency using Eq. 30.1.

f_L (calculated) = _____

b. Construct the circuit of Fig. 30.1. Apply an input of 1 V, rms. Vary the signal frequency from 100 Hz to 50 kHz while measuring and recording the output voltage in Table 30.1.

TABLE 30.1 Low-Pass Filter

f	100-Hz	500-Hz	1-kHz	2-kHz	5-kHz	10-kHz	15-kHz	20-kHz	50-kHz
V_o									

c. Plot the output gain-frequency response curve in Fig. 30.4.
d. Obtain the value of low-cutoff frequency from the data plotted in Fig. 30.4.

f_L (measured) = _____

Compare the low-cutoff frequency calculated in Part 1(**a**) with that obtained in Part 1(**d**).

Part 2. High-Pass Active Filter

a. Using Eq. 30.2, calculate the high-cutoff frequency for the circuit of Fig. 30.2.
b. Construct the circuit of Fig. 30.2. Apply an input of 1 V, rms. Vary the signal frequency from 1 kHz to 300 kHz and record the resulting output voltage in Table 30.2.

TABLE 30.2 High-Pass Filter

f	1-kHz	2-kHz	5-kHz	10-kHz	20-kHz	30-kHz	50-kHz	100-kHz	300-kHz
V_o									

c. Plot the data obtained in Table 30.2 in Fig. 30.5.
d. Using the plot in Fig. 30.5, obtain the high-cutoff frequency.

f_H (measured) = _____

Compare the high-cutoff frequency calculated in Part 2(**a**) with that measured in Part 2(**d**).

Exp. 30 / Procedure

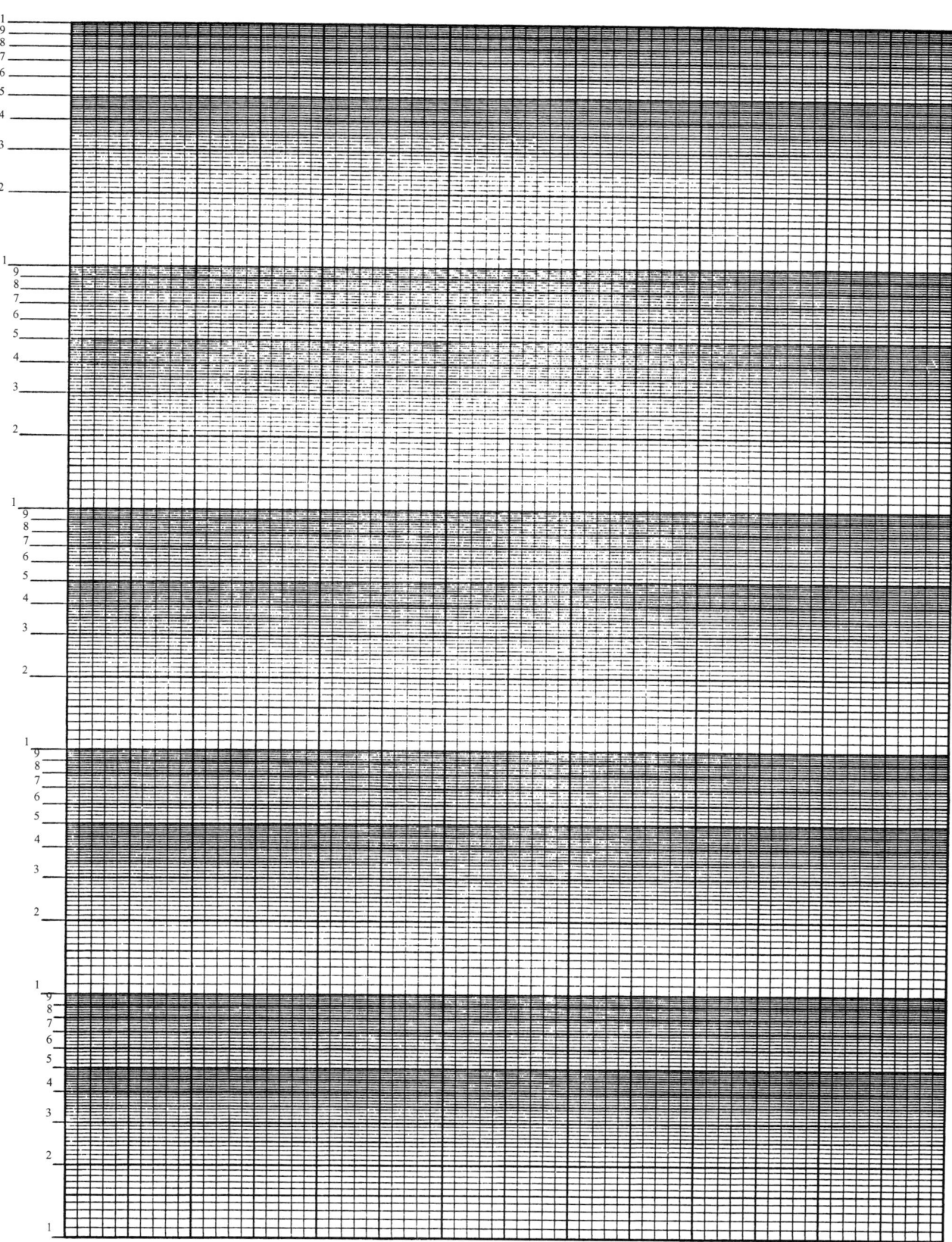

Figure 30-4

394 Exp. 30 / Active Filter Circuits

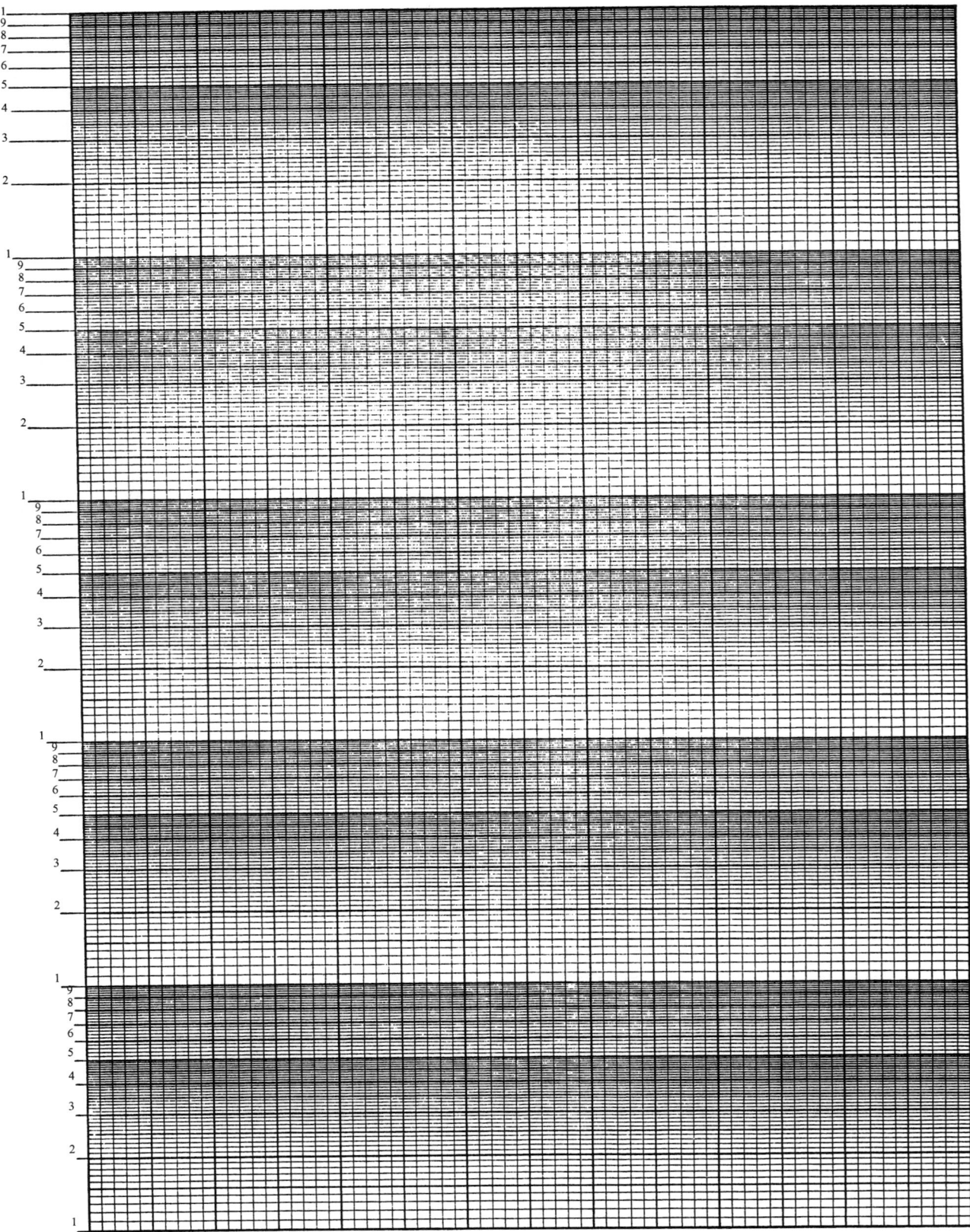

Figure 30-5

Exp. 30 / Procedure

Part 3. Band-Pass Active Filter

a. Calculate the band-pass frequencies using Eqs. 30.1 and 30.2.

b. Construct the circuit of Fig. 30.3.

c. Apply an input signal of 1 V, rms. Vary the signal frequency from 100 Hz to 300 kHz and record the output voltage in Table 30.3.

TABLE 30.3 Band-Pass Filter

f	100-Hz	500-Hz	1-kHz	2-kHz	5-kHz	10-kHz	15-kHz	20-kHz	30-kHz
V_o									

f	50-kHz	100-kHz	200-kHz	300-kHz
V_o				

d. Plot the data in Fig. 30.6. Using the plot, determine the lower and higher cutoff frequencies for the band-pass filter.

Compare the frequencies calculated in Part 3(a) with those measured in Part 3(d).

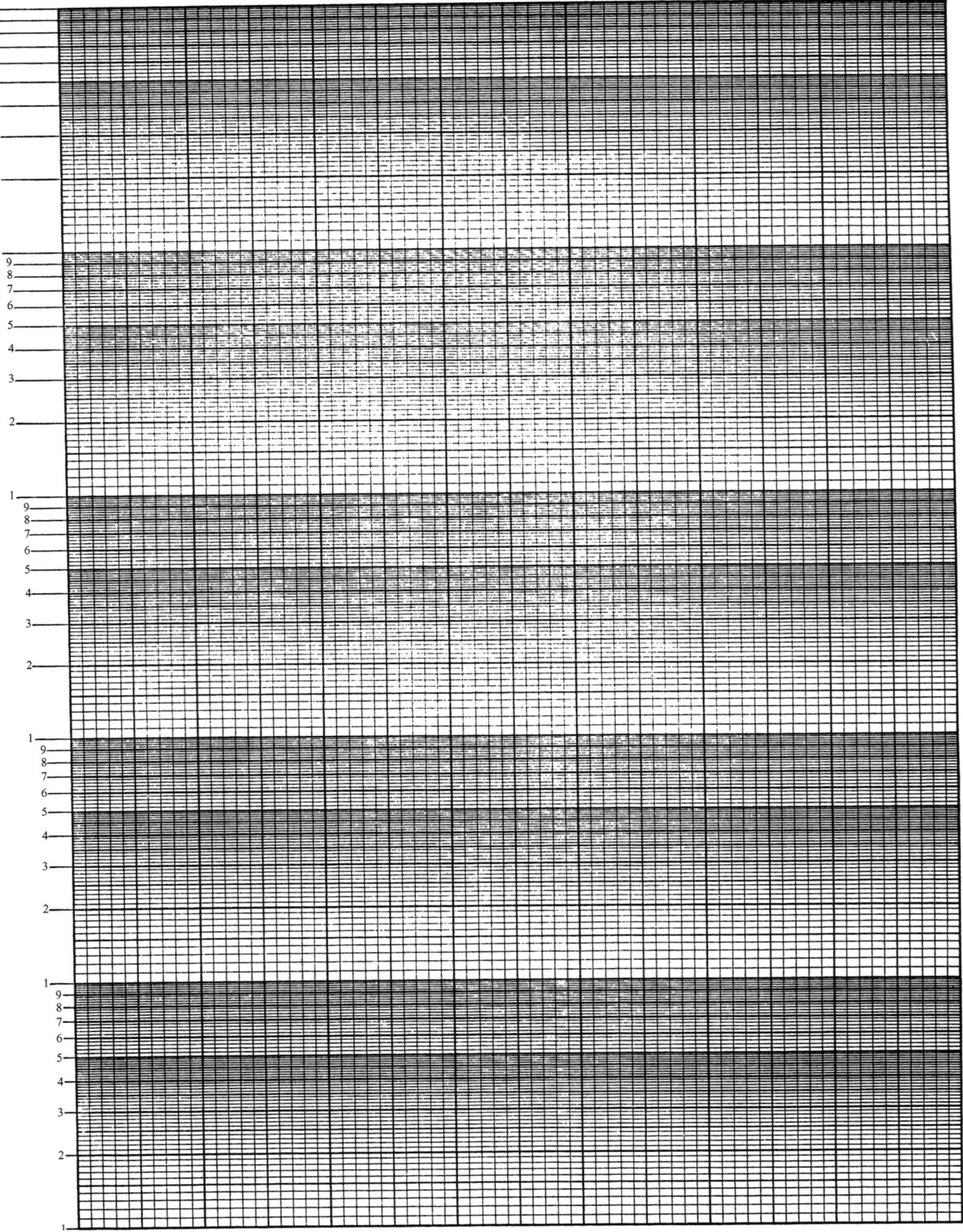

Figure 30-6

Part 4. Computer Exercises

PSpice Simulation 30-1

The circuit shown is that of the low-pass filter of Fig. 30.1.

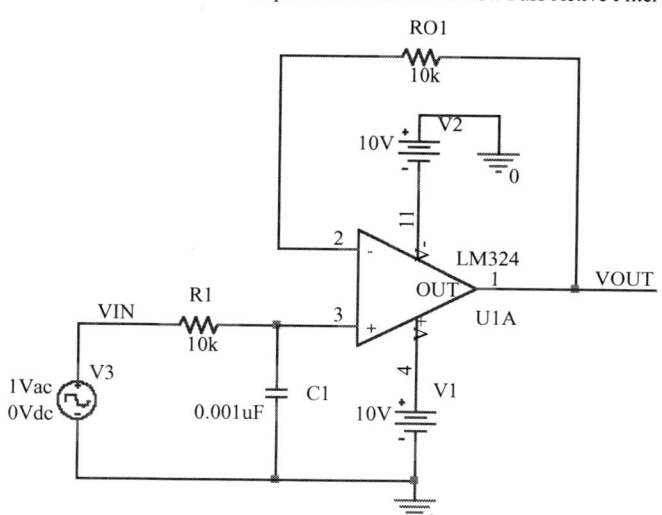

PSpice Simulation 30-1: Low Pass Active Filter

Perform an AC sweep (frequency) analysis for this circuit. Set the range for the sweep variable (frequency) from 1 Hz to 100 kHz. Select logarithmic sweep mode and specify 10 Points/Decade. Obtain the requested data and answer the questions:

1. Obtain a Probe plot of the voltage V(VOUT).

2. Use a cursor to obtain the cutoff frequency of this filter.

3. Obtain a Probe plot of DB(V(VOUT)).

4. Use a cursor to determine the –3dB frequency of that plot.

5. Compare the results of step 2 with those of step 4.

6. How do the results obtained compare to the low-cutoff frequency calculated from Eq. 30.1?

PSpice Simulation 30-2

The circuit shown is that of the band-pass filter of Fig. 30.3.

PSpice Simulation 30-2: Band-pass Active Filter

Perform an AC sweep (frequency) analysis for this circuit. Set the range for the sweep variable (frequency) from 100 Hz to 1.0 MHz. Select logarithmic sweep mode and specify 10 Points/Decade. Obtain the requested data and answer the questions:

1. Obtain a Probe plot of the voltage V(VOUT).

2. Use a cursor to obtain the center frequency of this filter.

Exp. 30 / Part 4. Computer Exercises

3. Compare this frequency to the cutoff frequencies of the low-pass and the high-pass filters.

4. Use the two PSpice cursors to determine the low- and high-cutoff frequencies.

5. From this data, determine the bandwidth of this filter.

6. Obtain a Probe plot of DB(V(VOUT)).

7. Use a cursor to determine the lower and upper −3dB frequencies of that plot.

8. From this data, determine the bandwidth of this filter.

9. Compare the results of step 4 with those of step 7.

10. Compare the results of step 5 with those of step 8.

Name _____
Date _____
Instructor _____

EXPERIMENT 31

Comparator Circuits Operation

OBJECTIVE

To measure DC and AC operation using comparator IC circuits.

EQUIPMENT REQUIRED

Instruments

Oscilloscope
DMM
Function generator
DC power supply

Components

Resistors

(1) 1-kΩ
(1) 3.3-kΩ
(3) 10-kΩ
(1) 20-kΩ
(3) 100-kΩ
(1) 50-kΩ potentiometer

Capacitors

(2) 15-μF
(1) 100-μF

Transistors and ICs

(1) 2N3904
(1) 741 op-amp IC (or equivalent)
(1) 339 comparator IC (or equivalent)
(1) LED (20 mA)

EQUIPMENT ISSUED

Item	Laboratory serial no.
DC power supply	
Function generator	
Oscilloscope	
DMM	

RÉSUMÉ OF THEORY

A comparator circuit is essentially a very high gain op-amp having a plus (+) and a minus (–) input. The output of the comparator is a logic level that provides an indication of when the plus input voltage is greater than the minus input or when the plus input is less than the minus input. Although an op-amp can be used for this purpose, special comparator ICs are available which are better suited for this operation.

Figure 31.1 shows a 741 op-amp used as a level detector. The reference level voltage V_{ref} is set at +5 V. The indicator LED goes *on* whenever the input V_i goes *below* V_{ref} and goes *off* whenever V_i goes *above* V_{ref}. Figure 31.2 shows a similar operation using a 339 comparator IC.

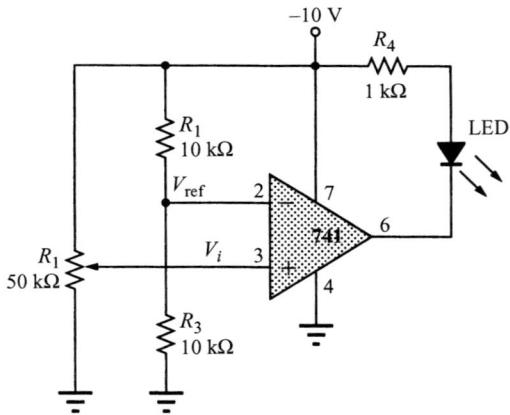

Figure 31-1

Figure 31.3 shows two comparator stages connected as a window detector—a circuit which indicates whenever the input voltage is *within* a specified range of voltage.

Exp. 31 / Procedure

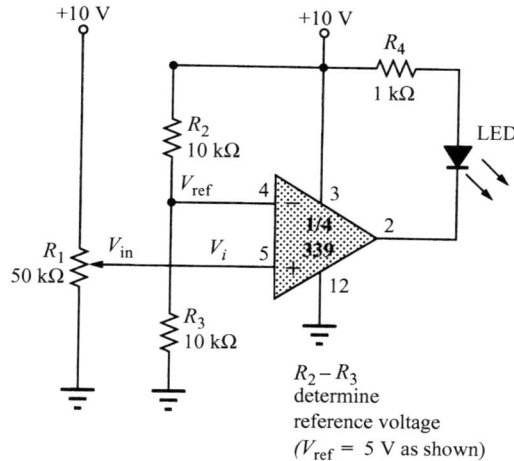

Figure 31-2

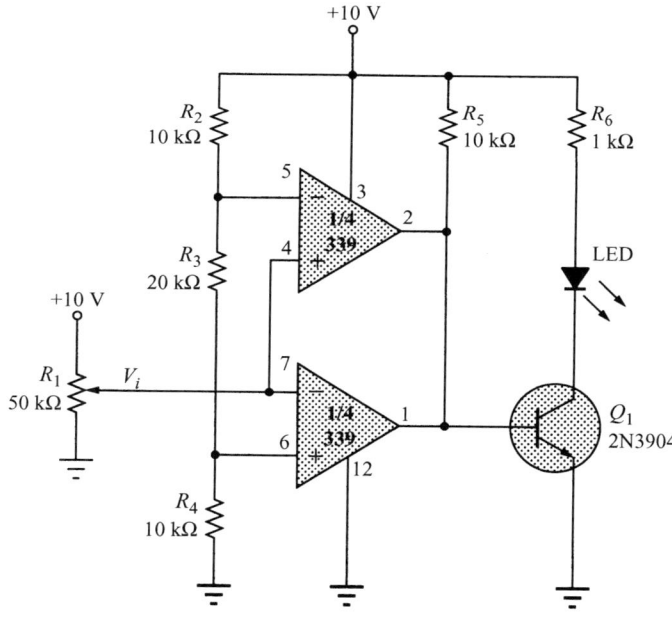

R_2, R_3, and R_4 set window voltages

Figure 31-3

PROCEDURE

Part 1. Comparator with 741IC Used as a Level Detector

a. For the circuit of Fig. 31.1 calculate V_{ref}.

V_{ref} (calculated) = _____ ($R_3 = 10$ kΩ)

b. Construct the circuit of Fig. 31.1. (Measure and record resistor values in Fig. 31.1.)

c. Using a DMM, measure the reference voltage, V_{ref}.

V_{ref} (measured) = _____

d. Adjust potentiometer R_1 so that the LED just goes *on*,* and then just goes *off*. Record the voltage V_i for each condition.

V_i (measured) (LED goes *on*) = _____
V_i (measured) (LED goes *off*) = _____

e. Replace R_3 with a 20-kΩ resistor and repeat Parts 1(**b**) and 1(**c**).

V_{ref} (measured) = _____
V_i (measured) (LED goes *on*) = _____
V_i (measured) (LED goes *off*) = _____

Compare the values of V_{ref} calculated in Part 1(**a**) with those measured in Parts 1(**c**) and 1(**e**).

---

*The potentiometer could be replaced by a rectified triangular wave at 100 Hz to provide the V_i input. The V_i signal and comparator V_o can then be observed simultaneously on the oscilloscope.

Exp. 31 / Procedure

Part 2. Comparator with ¼ 339IC Used as Level Detector

a. For the circuit of Fig. 31.2 calculate V_{ref}.

V_{ref} (calculated) = _____ (R_3 = 10 kΩ)

Repeat the calculation for R_3 = 20 kΩ.

V_{ref} (calculated) = _____ (R_3 = 20 kΩ)

b. Construct the circuit of Fig. 31.2. (Measure and record resistor values in Fig. 31.2.)

c. Using a DMM, measure the reference voltage.

V_{ref} (measured) = _____ (R_3 = 10 kΩ)

d. Adjust potentiometer R_1 so that the LED just goes *on** and also just goes *off*. Measure the input voltage for each condition.

V_i (measured) = _____ (LED goes *on*)

V_i (measured) = _____ (LED goes *off*)

e. Replace R_1 with a 20-kΩ resistor. Repeat steps c and d.

V_{ref} (measured) = _____ (R_3 = 20 kΩ)

V_i (measured) = _____ (LED goes *on*)

V_i (measured) = _____ (LED goes *off*)

---

*See footnote on page 404.

f. Interchange connections at pins 4 and 5 so that V_i goes to the *minus* input and V_{ref} goes to the *plus* input. Repeat Part 2(**d**).

V_i (measured) = _____ (LED goes *on*)

V_i (measured) = _____ (LED goes *off*)

Compare the calculated and measured voltages in Parts 2(**c**) through 2(**f**) with that calculated in Part 2(**a**).

Part 3. Window Comparator

a. For the circuit of Fig. 31.3 calculate V^+ (pin 5) and V^- (pin 6).

V^+ (calculated) (pin 5) = _____
V^- (calculated) (pin 6) = _____

b. Construct the circuit of Fig. 31.3. (Measure and record resistor values in Fig. 31.3.)

c. Using a DMM, measure the voltages at pins 1, 5, and 6.

V_i (measured) (pin 1) = _____
V^+ (measured) (pin 5) = _____
V^- (measured) (pin 6) = _____

Exp. 31 / Procedure

d. Adjust V_i from 0 V to +10 V.* Measure the voltage levels at which the LED goes *on* and then goes *off*.

V_i (measured) = _____ (LED goes *on*)

V_i (measured) = _____ (LED goes *off*)

e. Adjust V_i from +10 V to 0 V. Measure the voltage levels at which the LED goes *on* and then goes *off*.

V_i (measured) = _____ (LED goes *on*)

V_i (measured) = _____ (LED goes *off*)

f. Interchange resistors R_3 and R_4 and repeat Part 3(**d**).

V_i (measured) = _____ (LED goes *on*)

V_i (measured) = _____ (LED goes *off*)

Compare the calculated values in Part 3(**a**) with those measured in Part 3(**c**).

---

*See footnote on page 404.

Part 4. Computer Exercises

PSpice Simulation 31-1

The circuit shown is (almost) that of the comparator of Fig. 31.1.

PSpice Simulation 31-1: Comparator Circuit

The resistances of R3 and R4 shown are so chosen that VIN exceeds the voltage VREF. Run a bias point analysis and obtain the following data, and answer the stated questions:

1. Obtain all voltages and circuit currents.

2. Verify that VIN is greater than VREF.

3. If 8 milliamps is required to light up the diode D1, will the diode light up under the present operating condition of the circuit?

4. The resistance of R3 will now be changed to 6 kΩ, and that of R4 to 4 kΩ.

5. Repeat the bias point analysis.

6. Again, obtain all voltages and circuit currents.

7. Compare their values to those obtained from the previous analysis.

8. Verify that VIN is less than VREF.

9. Will the diode light up under the present operation condition of the circuit?

The careful reader will have noticed that VREF in this circuit is connected to the + input, while VIN is connected to the − input. This is different from Fig. 31.1. Compare the operation of this circuit with that of Fig. 31.1.

PSpice Simulation 31-2

The above circuit has been modified as shown.

PSpice Simulation 31-2: Comparator circuit with voltage pulse applied

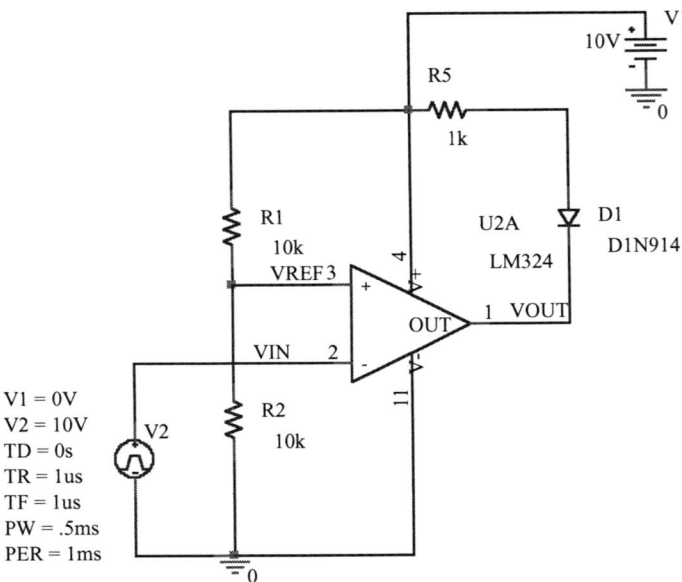

A voltage pulse source V2 with the parameters as shown has been applied to this circuit. The objective is to show that as VIN goes above the voltage VREF, a current will flow through the diode. When VIN goes below VREF, the diode current goes to near zero amps.

1. Run a 2-millisecond transient (time) analysis.

2. Plot V(VIN) and I(D1) on two different *y*-axes. This is necessary because of their different amplitudes.

3. Explain your findings. Has the objective been obtained?

EXPERIMENT 32

Oscillator Circuits 1: The Phase-Shift Oscillator

Name _____
Date _____
Instructor _____

OBJECTIVES

1. To measure the amplitude, the frequency of Vout, and the open-loop gain of the op-amp of the phase-shift oscillator.
2. To compare these experimental results with their theoretical values.
3. If a deviation of more than 10% between these two sets of data is obtained, suggest and test a modification of the circuit to bring the experimental results to within that deviation.
4. Run a PSpice analysis of this circuit and obtain the amplitude, the frequency of Vout, and the phase shift of the feedback network. Observe the effect on Vout when the saturation voltages V2 and V3 are reduced. Compare the PSpice data with the experimental data.

EQUIPMENT REQUIRED

Instruments

Oscilloscope (dual trace)
DC power supply

Components

Capacitors

(3) 0.1-µF

Resistors

(3) 1-kΩ
(1) 27-kΩ
(1) 1-MΩ
(1) 0-5-kΩ potentiometer

IC
(1) 741 (or equivalent) op-amp

EQUIPMENT ISSUED

Item	Laboratory serial no.
Oscilloscope(dual trace)	
DC power supply	

RÉSUMÉ OF THEORY

The circuit used in this experiment will exhibit sinusoidal oscillations at a particular frequency, defined as the oscillating frequency, without an external signal applied. For this to occur, two conditions must be met.

The first is that the closed-loop gain computed as the product of the gain of the op-amp multiplied by the gain of the feedback network of the circuit must be numerically equal to unity.
The gain of the feedback network at the oscillating frequency is equal to 1/32. It follows that the gain of the op-amp must be approximately equal to or greater than 32 to achieve a unity gain for the closed loop. The needed op-amp gain is computed as follows:

$$\frac{(RPot + Rf)}{Rin} \geq 29$$

The second condition for oscillation to occur is that the phase shift around the closed loop consisting of the op-amp and the three RC sections is 0°. The feedback network provides an overall phase shift of 180° and the op-amp will provide a phase shift of −180° at the oscillating frequency. Its value is computed as follows:

$$f = \frac{1}{\left(2\pi RC\sqrt{6}\right)} Hz$$

PROCEDURE

Part 1. Determining Vout

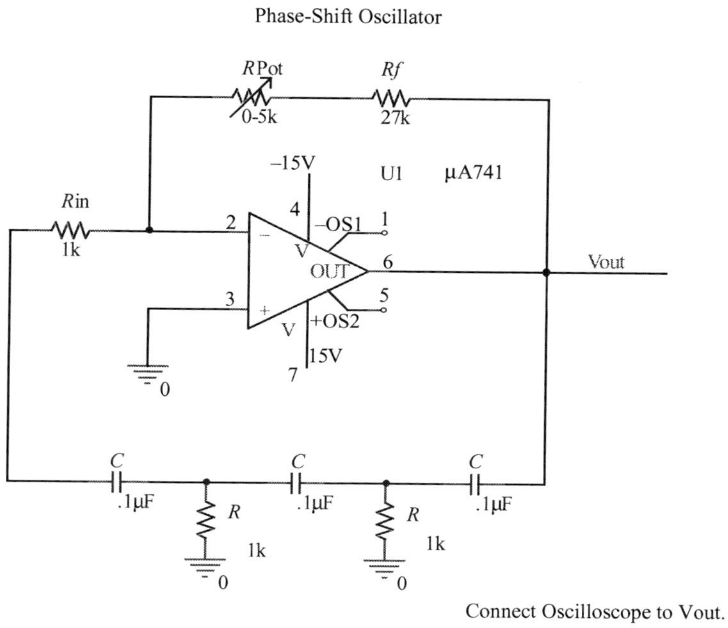

Phase-Shift Oscillator

Figure 32-1

Connect Oscilloscope to Vout.

a. Wire the circuit shown in Fig. 32-1.

b. Connect terminal 7 of the op-amp to the positive terminal of the power supply. Set the voltage of that terminal to 15 V. Connect terminal 4 of the op-amp to the negative terminal of the power supply. Set the voltage of that terminal to −15 V.

c. Connect the oscilloscope to the Vout terminal of the phase-shift oscillator.

d. Using the resistance value of R and the capacitance value of C, calculate the theoretical frequency of Vout using the formula given on the previous page.

f (theoretical) = _____Hz

e. Adjust the time base of the oscilloscope to obtain several cycles of Vout. Adjust the vertical scale to about 5 V/division. Select AC coupling.

f. Estimate the approximate resistance setting of RPot for oscillations to occur. Remember that the gain of the op-amp must be approximately 32.

Estimated Setting of RPot = _____ kΩ

g. Carefully adjust the resistance of RPot to obtain oscillations. Note: if you increase the resistance of RPot, eventually the output sine wave will show clipping. Adjust the resistance of RPot until the oscillator just sustains oscillations. Record the peak-peak voltage Vout.

Vout(peak-peak) = _____ V

h. Change the time base so that about two cycles of Vout are displayed. Measure the period of one of these cycles.

Period = _____ ms

i. From the above value for the period of Vout, calculate the value of the experimental frequency obtained.

f(experimental) = _____ Hz

j. Obtain the percent difference between the theoretical and the experimental frequencies using the theoretical frequency as the standard of comparison.

$$\% \text{ difference} = \frac{f(\text{theoretical}) - f(\text{experimental})}{f(\text{theoretical})} \cdot 100$$

Calculated % difference = _____

k. Disconnect RPot and Rf from the circuit. Use a DMM and measure their combined resistance.

RPot + Rf = _____ Ω

l. Using this value of the resistance and assuming that Rin is 1kΩ, calculate the open-loop gain for this oscillator.

Open-loop gain = _____

m. Obtain the percent difference between the theoretical gain of 32 and the experimental gain of the op-amp.

Calculated % difference = _____

n. If the experimental frequency and the open-loop gain of the op-amp differ by more than 10% from their theoretical values, suggest a change of the design to bring these two sets of data within that percent deviation.

Part 2. Computer Exercises

PSpice Simulation

a. Modify the circuit of Fig. 32-1 as shown in Figure 32.2. Terminals 4 and 7 of the op-amp remain connected to their respective voltages as in Fig. 32-1. The resistors and the capacitors each need a unique name in PSpice. The thermal variations in the physical circuit which start the oscillation process have been modeled by the voltage source VSRC. This allows us to run a Time Domain (transient) analysis of the circuit. From it we can determine the amplitude and the frequency of Vout. The VSRC source also allows us to run an AC Sweep/Noise (frequency) analysis. This will allow us to determine the phase angle between the voltages Vout and Vfeedback.

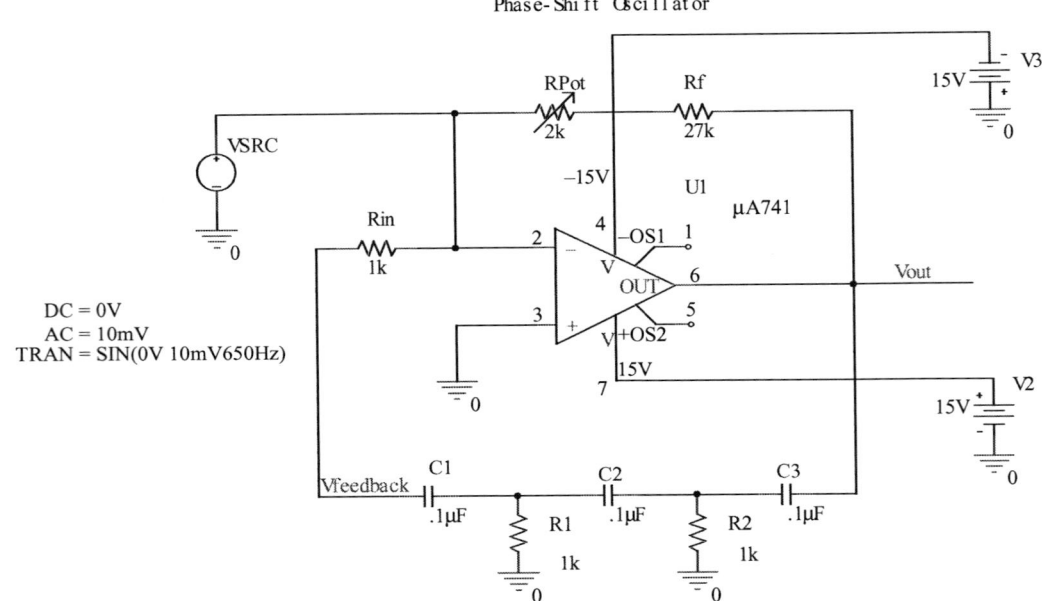

Connect Oscilloscope to Vout.

Figure 32-2

b. Run a Time Domain (transient) analysis of this circuit. In the "Simulation Setting box" set the Run time to 5 ms. Obtain a probe plot of Vout and record its peak-peak value.

Vout(peak-peak) = _____ V

Exp. 32 / Part 2. Computer Exercises

c. On a probe plot, measure the period of Vout using the two probe cursors.

Vout(period) = _____ ms

d. Calculate the frequency of Vout from the previous data.

Vout(frequency) = _____ Hz

e. Reduce the saturation voltages V2 and V3 to 10V and –10V, respectively, and repeat the analysis. Record the peak-peak voltage of Vout. What is the effect on Vout?

Vout(peak-peak) = _____ V

f. Measure the period of Vout on a probe plot and from it calculate its frequency. Compare this with its value obtained in **d.** above. What is your conclusion?

Vout(frequency) = _____ Hz

g. Compare the PSpice data obtained with the experimental data. Are they in agreement? If not, can you explain the discrepancy?

h. Run an AC Sweep/Noise(frequency) analysis of the circuit. Set the Start frequency to 1 Hz and the End frequency to 2000 Hz. Set the logarithmic sweep at 10 points/decade.

i. On a Probe plot, obtain traces of $P(V(\text{feedback}))$ and $P(V(\text{VOUT}))$. Remember to select P for phase in the Macro or Function box. This will plot the phase of the above two voltages. Select the former trace and set cursor A1 to 650 Hz. Read the corresponding phase shift. Select the latter trace, the cursor A1 remaining at 650 Hz. Read the corresponding phase shift.

j. What is the phase difference between these two voltages? Does the data obtained conform to expectation?

Name
Date
Instructor

EXPERIMENT 33

Oscillator Circuits 2

OBJECTIVE

To generate and measure voltage waveforms in various oscillator circuits.

EQUIPMENT REQUIRED

Instruments

Oscilloscope
DMM
Function generator
DC power supply

Components

Resistors

(3) 10-kΩ
(1) 51-kΩ
(1) 100-kΩ
(1) 220-kΩ
(1) 500-kΩ pot

Capacitors

(3) 0.001-µF
(3) 0.01-µF
(1) 15-µF

ICs

(1) 7414 Schmitt-trigger IC
(1) 741 (or equivalent) op-amp
(1) 555 timer IC

EQUIPMENT ISSUED

Item	Laboratory serial no.
DC power supply	
Function generator	
Oscilloscope	
DMM	

RESUMÉ OF THEORY

Wien-Bridge Oscillator

A bridge network can be used to provide the 180° phase shift as shown in Fig. 33.1. The circuit's resulting frequency can be calculated from

$$f = \frac{1}{2\pi \sqrt{R_1 C_1 R_2 C_2}} \tag{33.1}$$

If $R_1 = R_2 = R$, and $C_1 = C_2 = C$, then

$$f = \frac{1}{2\pi RC} \text{ Hz} \tag{33.2}$$

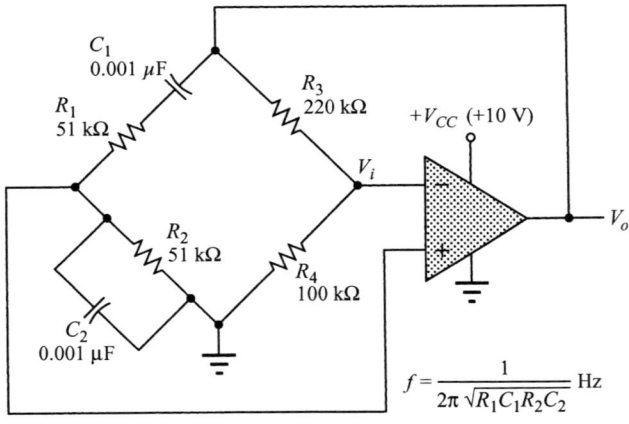

Figure 33-1

Square-Wave Oscillator

A 555 timer IC is a versatile linear digital IC which can be wired for operation as an oscillator, as shown in Fig. 33.2. The output resulting from this circuit is a pulse clock waveform of frequency

$$f = \frac{1.44}{(R_A + 2R_B)C} \text{ Hz} \tag{33.3}$$

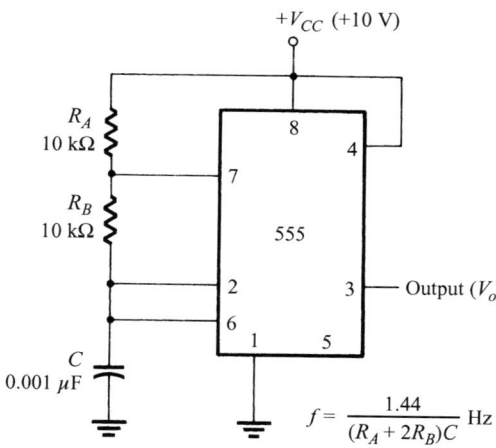

Figure 33-2

Schmitt-Trigger Oscillator

A single Schmitt-trigger IC, resistor, and capacitor can be used to build a pulse-type oscillator circuit, as shown in Fig. 33.3. The oscillator frequency is generally calculated using

$$f = \frac{k}{RC} \text{ Hz} \qquad (33.4)$$

where k is typically 0.3 to 0.7, depending on the internal triggering levels of the Schmitt-trigger IC.

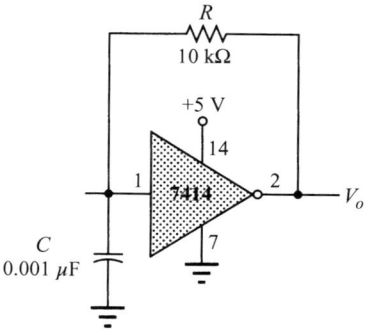

Figure 33-3

PROCEDURE

Part 1. Wien Bridge Oscillator

a. Construct the circuit of Fig. 33.1. (Measure and record the resistor values in Fig. 33.1.)

b. Using the oscilloscope, observe and record the output waveform in Fig. 33.4.

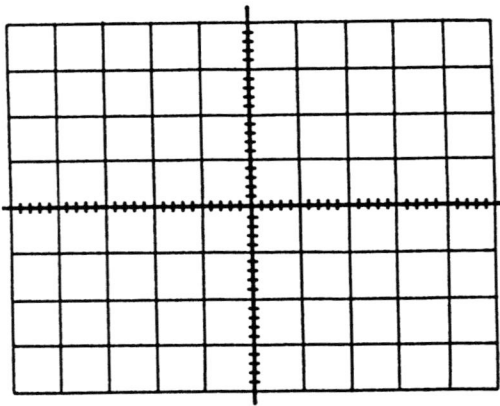

Figure 33-4

c. Measure the time for one cycle.

T (measured) = _____

d. Calculate the signal frequency.

$f = 1/T =$ _____

e. Change both capacitors to $C = 0.01$ µF and repeat Parts 2(**c**) through 2(**d**).

T (measured) = _____
$f = 1/T =$ _____

f. Calculate the theoretical frequency of the oscillator for each capacitor value.

$$f(C = 0.001~\mu F)~\text{(calculated)} = \underline{\hspace{2cm}}$$
$$f(C = 0.01~\mu F)~\text{(calculated)} = \underline{\hspace{2cm}}$$

Compare the calculated frequencies for both capacitor values with those measured.

Part 2. 555 Timer Oscillator

a. Construct the oscillator circuit of Fig. 33.2. (Measure and record the resistor values in Fig. 33.2.)

b. Observe and record the output waveforms at pins 3 and 4 in Fig. 33.5.

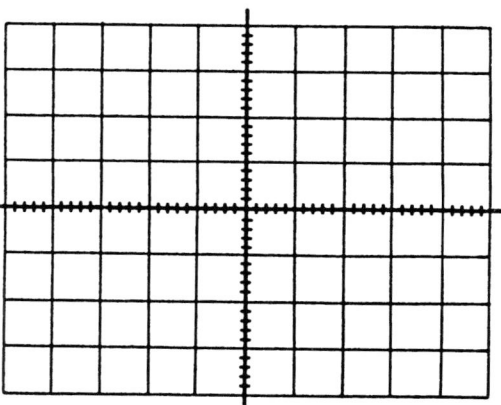

Figure 33-5

c. Measure the period of the output waveform.

T (measured) = \underline{\hspace{2cm}}

d. Calculate the signal frequency.

$$f = 1/T = \underline{\hspace{2cm}}$$

e. Replace the capacitor with $C = 0.01$ µF, and repeat Parts 3(**c**) through 3(**d**).

$$T \text{ (measured)} = \underline{\hspace{2cm}}$$
$$f = 1/T = \underline{\hspace{2cm}}$$

Part 3. Schmitt-Trigger Oscillator

a. Construct the oscillator circuit of Fig. 33.3. (Measure and record the resistor value in Fig. 33.3.)

b. Observe and record the output waveforms at pins 1 and 2 in Fig. 33.6.

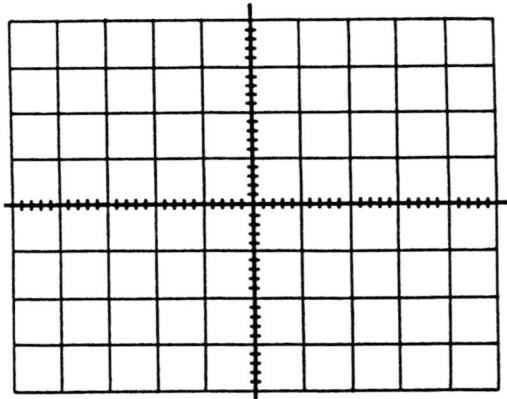

Figure 33-6

c. Measure the period of the output waveform.

T (measured) = _____

d. Calculate the signal frequency.

$f = 1/T =$ _____

e. Replace the capacitor with $C = 0.01$ μF, and repeat Parts 4(**c**) and 4(**d**).

T (measured) = _____
$f = 1/T =$ _____

f. Calculate the theoretical frequency, f, using Eq. 33.4 for each of the capacitor values.

$f(C = 0.001$ μF$)$ (calculated) = _____
$f(C = 0.01$ μF$)$ (calculated) = _____

Compare the calculated value of f for each capacitor with those measured.

Part 4. Computer Exercises

PSpice Simulation 33-1

The circuit shown is very similar to that of Fig. 33.2. Note that in the circuit shown below, terminal 5 of the 555 timer has been connected to a voltage divider consisting of the two resistors R1 and R2.

PSpice Simulation 33-1: Square Wave Oscillator

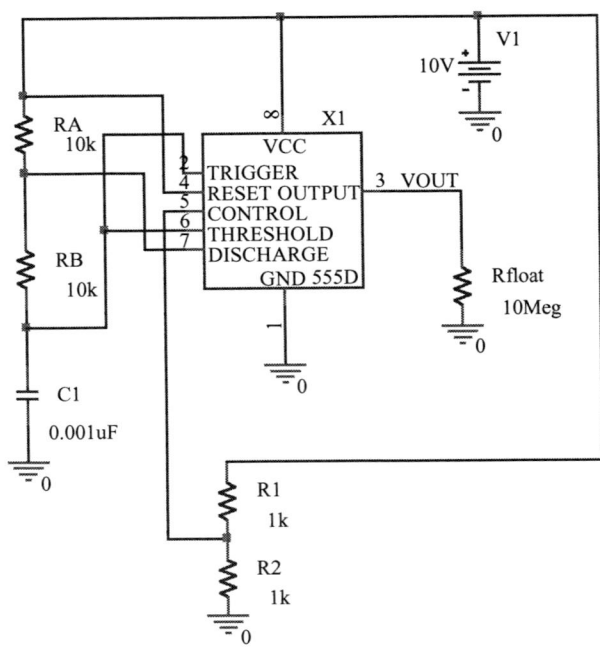

For this circuit, run a 100-microsecond transient (time) analysis. Obtain the requested data and answer the questions:

1. With the values of R1 and R2 both at 1 kΩ, what is the minimum amplitude of VOUT and what is its maximum amplitude?

2. Is VOUT a square pulse?

3. What is the period of VOUT?

4. What is the pulse width of VOUT?

5. What is the fundamental frequency of VOUT?

Exp. 33 / Part 4. Computer Exercises

6. Change R2 to 5 kΩ and repeat the transient analysis.

7. Is VOUT still a square pulse?

8. Have its amplitudes changed from the previous analysis?

9. What is the new period of VOUT?

10. What is the new pulse width of VOUT?

11. What is the new fundamental frequency of VOUT?

12. This circuit, with its control terminal connected as shown, is defined as a voltage-controlled oscillator (VCO). Does your data confirm this?

Name _____
Date _____
Instructor _____

EXPERIMENT 34

Voltage Regulation— Power Supplies

OBJECTIVE

To measure DC and ripple voltages in series and shunt regulator circuits.

EQUIPMENT REQUIRED

Instruments

Oscilloscope
DMM
Function generator
DC power supply

Components

Resistors

(1) 390-Ω, 2-W
(2) 1-kΩ
(1) 2-kΩ
(1) 20-kΩ

Transistors and ICs

(1) *npn* power transistor
(1) op-amp (741 or equivalent)

Exp. 34 / Voltage Regulation—Power Supplies

EQUIPMENT ISSUED

Item	Laboratory serial no.
DC power supply	
Function generator	
Oscilloscope	
DMM	

RÉSUMÉ OF THEORY

Voltage regulators attempt to maintain a constant DC output voltage by controlling the series current fed to the load—series voltage regulation, or by controlling the current to the load by shunting some of it away.

Series Regulation

The circuit of Fig. 34.1 shows a basic series regulator circuit. The Zener diode provides a reference voltage which sets the output voltage at

$$V_L = V_Z - V_{BE} \tag{34.1}$$

If the output voltage tends to go lower, the series transistor is driven further into conduction, providing more current to the load to maintain the output voltage.

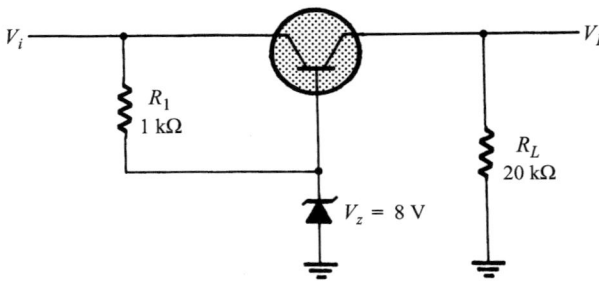

Figure 34-1

Improved Series Regulation

The circuit of Fig. 34.2 shows the addition of an op-amp to provide improved regulation. The output voltage is set by the Zener diode and feedback network made of resistors R_1 and R_2. The voltage gain of the op-amp, connected in a positive-feedback configuration, is

$$A = \frac{R_1}{R_2} \tag{34.2}$$

with

$$V_L = A V_Z \tag{34.3}$$

If the output voltage tends to get larger, the increased feedback voltage sensed by voltage divider R_1 and R_2 causes a reduced input to the op-amp, less drive current to the series pass transistor, and reduced load current, thereby maintaining the output voltage.

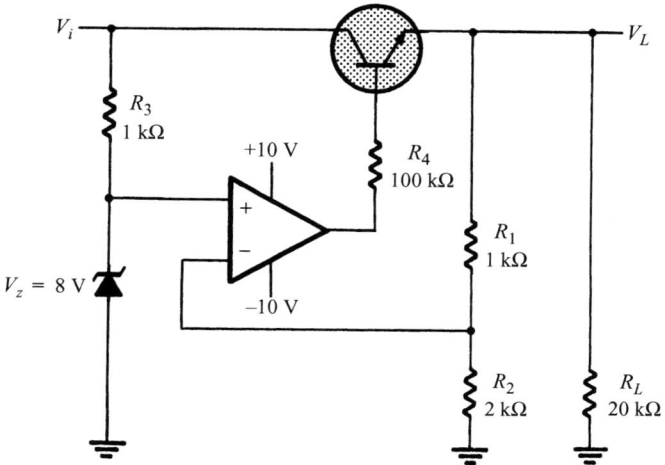

Figure 34-2

Shunt Regulation

The circuit of Fig. 34.3 shows a transistor connected in parallel (shunt) with the output. The transistor conducts to provide greater or less load current, thereby maintaining the output voltage. Again, a sensing network made of a resistor voltage divider (using R_1 and R_2) controls the input to the op-amp, which then controls the conduction of the shunt transistor. The regulated output voltage can be calculated using

$$V_L = \frac{R_1 + R_2}{R_1} V_Z \tag{34.4}$$

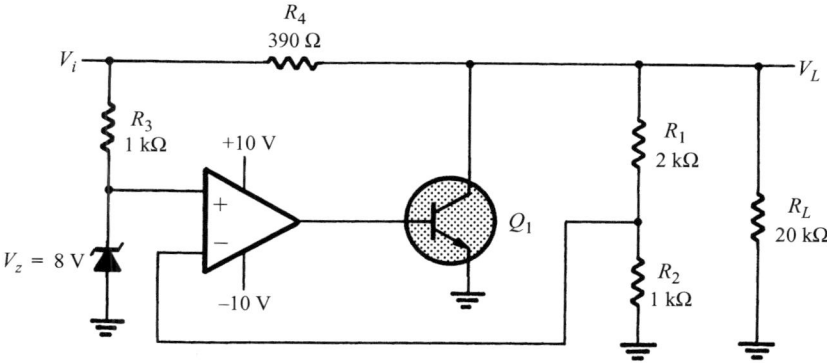

Figure 34-3

PROCEDURE

Part 1. Series Voltage Regulator

a. Calculate the resulting regulated voltage for the circuit of Fig. 34.1.

b. Construct the circuit of Fig. 34.1. (Measure and record the resistor values in Fig. 34.1.) Vary the DC input voltage, V_i, from 10 V to 16 V, measuring and recording the load voltage in Table 34.1. Record the regulated output voltage measured.

V_o (measured) = _____

TABLE 34.1 Series Voltage Regulator

V_i	10-V	11-V	12-V	13-V	14-V	15-V	16-V
V_o							

Compare the regulation voltage obtained in Part 1(**b**) with that calculated in Part 1(**a**).

Part 2. Improved Series Regulator

a. Calculate the regulated output voltage for the circuit of Fig. 34.2.

V_L (calculated) = _____

b. Construct the circuit of Fig. 34.2. (Measure and record the resistor values in Fig. 34.2.) Vary the DC input voltage, V_i, from 10 V to 24 V, in 2-V steps, measuring and recording the load voltage, V_L, in Table 34.2. Record the value of the regulated load voltage.

V_L (measured) = _____

TABLE 34.2 Series Voltage Regulator

V_i	10-V	12-V	14-V	16-V	18-V	20-V	22-V	24-V
V_L								

Exp. 34 / Procedure

Compare the regulation voltage obtained in Part 2(**b**) with that calculated in Part 2(**a**).

Part 3. Shunt Voltage Regulator

a. Calculate the regulated voltage from the circuit of Fig. 34.3.

V_L (calculated) = _____

b. Construct the circuit of Fig. 34.3. (Measure and record the resistor values in Fig. 34.3.) Apply an input voltage varied from 24 V to 36 V, in 2-V steps. Measure and record the load voltage in Table 34.3. Record the regulated output voltage.

V_L (measured) = _____

TABLE 34.3 Series Voltage Regulator

V_i	24-V	26-V	28-V	30-V	32-V	34-V	36-V
V_o							

Compare the regulation voltage obtained in Part 3(**b**) with that calculated in Part 3(**a**).

Part 4. Computer Exercises

PSpice Simulation 34-1

The circuit shown is almost identical with that of the series voltage regulator of Fig. 34.1. In the present case, however, the Zener voltage VZ is at 4.68 V. Perform a DC sweep analysis in which the voltage of source V1 changes from 2 V to 8 V in steps of 0.5 V.

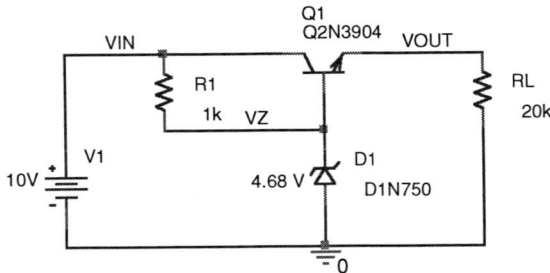

PSpice Simulation 34-1: Series Voltage Regulator

Obtain the requested data and answer all questions:

1. Obtain Probe plots of V(VIN), V(VZ), and V(VOUT).

2. Explain the plot of V(VIN).

3. What are the voltages of V(VZ) and V(VOUT) when V(VIN) is equal to 8 V?

4. At what voltage of V(VIN) do V(VZ) and V(OUT) attain their steady-state voltages?

5. What is the voltage difference between the voltages V(VZ) and V(VOUT)?

6. Is this voltage equal to the base-to-emitter voltage of Q1?

7. From the obtained data, verify Eq. 34.1.

PSpice Simulation 34-2

The circuit shown is that of the shunt regulator of Fig. 34.3. Again, the Zener potential is 4.68 V. Perform a DC sweep analysis in which the voltage of source V3 changes from 6 V to 16 V in steps of 0.5 V.

PSpice Simulation 34-2: Shunt Regulation

Obtain the requested data and answer all questions:

1. Obtain Probe plots of V(VIN), V(VZ), and V(OUT).

2. Describe the appearance of the plot of V(VIN).

3. Calculate the theoretical value of V(OUT) using Eq. 34.4.

4. Compare that value with the value obtained from the Probe data.

5. What are the final voltages of V(VZ) and V(VOUT) when V(VIN) is equal to 16 V?

Name _____
Date _____
Instructor _____

EXPERIMENT 35

Analysis of AND, NAND, and INVERTER Logic Gates

OBJECTIVES

1. To analyze logical primitives using the **PSpice** program.
2. To construct circuits in the laboratory using these logical primitives.
3. To compare the results obtained in the laboratory with those obtained from simulation.
4. To make any needed corrections in the circuits.

EQUIPMENT REQUIRED

Instruments

Computer able to run **OrCad PSpice**, version 9.1 or higher
Cadet logic trainer (or equivalent)
Tektronic 2236 oscilloscope (or equivalent)
10:1 scope probe
DMM
7400, 7404, 7408 gates

EQUIPMENT ISSUED

Item	Laboratory serial no.
Logic trainer	
Oscilloscope	
Scope probe	

RÉSUMÉ OF THEORY

All digital operations depend on three fundamental logic operations: the **AND, OR,** and **NOT-AND**. All logical operations, no matter how complex, can be reduced to these three basic functions. In this laboratory exercise, we will study the digital devices that perform these functions. They are known as *gates*. In their simplest form, they are also known as *logical primitives*. The first we will study is the **AND** gate.

PROCEDURE

For this and subsequent experiments, the student is to simulate all logical circuits using the **PSpice** program before he/she performs the actual experiment, which conforms to engineering practice. The result of the simulation can then be discussed with the laboratory instructor, preferably in the week prior to a particular experiment.

After completion of the simulation, the design is taken into the laboratory. After the actual circuit is constructed and measurements are taken, the student should compare the results obtained from the simulation with those obtained in the laboratory. In this way, any discrepancy between the results can be cleared up.

The laboratory procedures are stated in a general fashion because the authors do not know what particular laboratory equipment is available to the students. Hence, the experimental steps in a laboratory may have to be somewhat modified from those specified.

Part 1. AND Gate: Computer Simulation

Figure 35.1 shows a two-input **AND** gate. The **AND** gate is located in the **eval.slb** library of the **PSpice** program. Its shape tells its logical function. The circuit consists of the 7408 logic gate. Digital stimulus generators, **DSTM1** and **DSTM2,** are connected to the input terminals 1 and 2 of the gate, respectively. They are located in the **Source** library of the **PSpice** program. Terminal 3 is the output terminal.

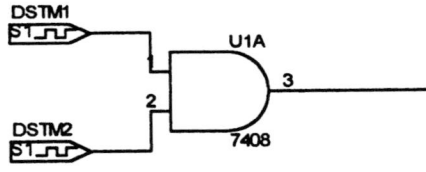

Figure 35-1

Our objective is to obtain the logic function of the **AND** gate.

Each of the stimulus generators can assume two states. Logically they are known as 1 or **ON**, or 0 or **OFF**. Thus, there are four logic states at the input terminals. The **AND** gate will respond to each of those states. The conditions specified at the input terminals are logic states, not electrical quantities such as voltage. To keep track of the possible logic states, construct Table 35.1.

TABLE 35.1

Input terminal 1	Input terminal 2	Output terminal 3
1	1	
0	1	
1	0	
0	0	

In this table, all possible logic states of the input terminals are present.

Next, run a 500-millisecond **PSpice** simulation to obtain the logic states at Output terminal 3. The logic states of **DSTM1** and **DSTM2** are set using the **OrCad Capture-[Property Editor]**.*

Set the logic states at the input terminals as shown on the **PROBE** plot (Fig. 35.2). The trace of Input terminal 1 is shown as **U1A:A**. Its frequency is set at 10 Hz. The trace of Input terminal 2 is shown as **U1A:B**. Its frequency is set at 5 Hz. These frequencies are set relatively low to allow the experimenter to observe the logic states if an **LED** is used during the experiment.

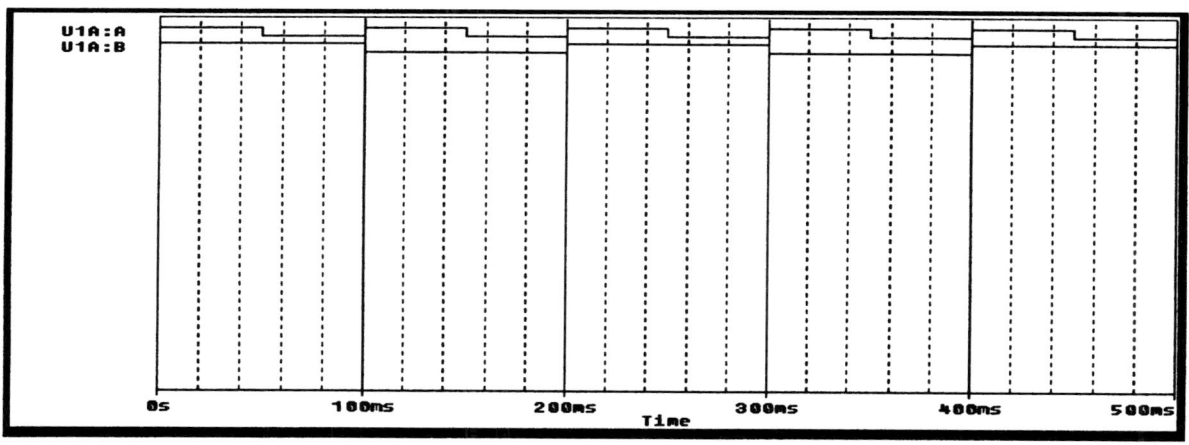

Figure 35-2

The **PROBE** plot shows that the logic states of the input terminals are identical to those listed in Table 35.1. After the logic states of the input terminals are set, perform the simulation of Fig. 35.1. In addition to the two traces in the above **PROBE** plot, now obtain the logic state of Output terminal 3. Its trace is shown as **U1A:Y**. The letters **A** and **B** are always used to define input terminals. The letter **Y** is used to define the output terminal.

---

*See Franz J. Monssen, *OrCad PSpice with Circuit Analysis*, 3rd Ed. (Upper Saddle River, NJ: Prentice Hall, 2001), p. 23.

From the simulation data obtained, answer the following:

   a. Complete the logic states for Output terminal 3 in Table 35.1.

   b. What logical relationship exists between the input terminals and the output terminal?

   c. What is the most common logic state of Output terminal 3?

   d. What are the logic states of the 7408 gate at 25 ms, 125 ms, and 375 ms? *Hint:* Move a **PROBE** cursor to the indicated times. The logic state of each terminal will appear on the left margin of the **PROBE** plot.

Part 2. AND Gate: Experimental Determination of Logic States

This determination is performed using either the Cadet Trainer of E & L Instruments (or its equivalent) and the Tektronic 2236 oscilloscope.

Proceed as follows:

1. Plug the 7408 gate into the protoboard of the Cadet Trainer.
2. Wire pin 14 to +5 volts and pin 7 to ground (zero volts).
3. Connect the **TTL** clock of the trainer to Input terminal 1. This provides a train of pulses. Their fundamental frequency can be set.
4. Set the fundamental frequency of the **TTL** clock to 10 Hz.
5. Connect Input terminal 2 of the gate to the **HI/LO** logic switch of the trainer.
6. Connect Output terminal 3 of the gate to a logic monitor or the logic probe. It can also be wired to an LED of the logic trainer. This gives a direct visual indication of the **ON** or **OFF** state of the **AND** gate.
7. Connect channel 1 of the scope to monitor the **TTL** clock.
8. Connect channel 2 of the scope to monitor at Output terminal 3.

Exp. 35 / Procedure

9. Turn on the power to the trainer.
10. Toggle the **HI/LO** logic switch to obtain the logic state of the gate.

Note: Whenever the **TTL** clock input and the **HI/LO** switch are high, Output terminal 3 will be high. When either or both are low, there will be no output present. Toggle the **HI/LO** switch to obtain the logic conditions for the gate. At this low frequency, it is best to set the scope trigger to **chop**, not **alternate**.

From the experimental data obtained, answer the following questions:

    a. How does the period of the pulse of the **TTL** pulse train compare to the period of the **PROBE** trace of **U1A:A**?

    b. What is the frequency of that trace and that of the **TTL** pulses?

    c. Are the logic states of the 7408 the same in both the computer simulation and in the experiment as measured at 25 ms, 125 ms, and 375 ms?

    d. What was the voltage amplitude of the **TTL** pulses and that of Output terminal 3?

    e. If there was a difference in their respective voltages, how can this be explained?

Part 3. Logic States versus Voltage Levels

The **PSpice Probe** plot provided traces of the logic states of the three terminals of the 7408 gate. The experimental data provided two distinct voltage levels at these terminals corresponding to the logic states. The high voltage output of the 7408 gate corresponded to a logical 1, or **ON**, on the **PROBE** plot. The low voltage output of that gate corresponded to a logical 0, or **OFF**, on the **PROBE** plot. We can simulate the experimental voltage levels using the **PSpice** program.

Exp. 35 / Analysis of AND, NAND, and INVERTER Logic Gates

First, modify the logic circuit of Fig. 35.1, shown in Fig. 35.3.

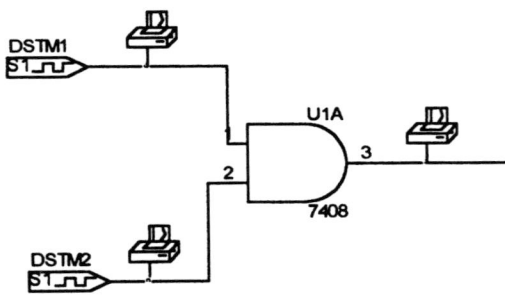

Figure 35-3

To each terminal, a **VPLOT1** device is attached. It can be found in the **Special** library of the **PSpice** program. Run a simulation.

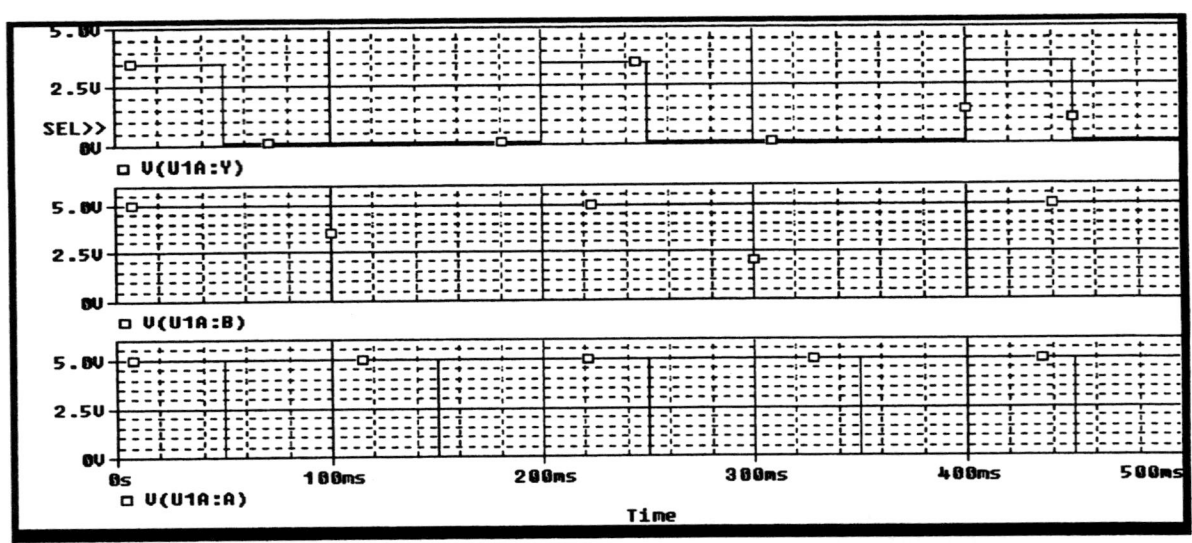

Figure 35-4

The results are shown on the **PROBE** plot in Fig. 35.4. It shows that the two input voltages are 5 volts at maximum. The output voltage has a maximum value of 3.5 volts. The traces now have the letter **V** (for voltage) attached to them. The terminal designations **A**, **B**, and **Y** are retained.

a. Run a **PSpice** simulation to verify this result.

b. Calculate the percent deviation between the experimentally obtained output voltage VY and the simulated output voltage shown on the **PROBE** plot as trace V(V1A:Y). Use the following formula:

$$\% \text{ deviation} = \frac{V(V1A:Y) - VY}{V(V1A:Y)} * 100\%$$

% deviation = _____

In this calculation, V(V1A:Y) is the standard of comparison.

c. If a 10% deviation between the simulated voltage V(V1A:Y) and the experimental voltage VY is acceptable, does the calculated percent deviation fall within the permissible range?

Part 4. Propagation Delay

The propagation delay of a gate is defined as the time interval between the leading edge of the input pulse and the leading edge of the output pulse as both have reached 50% of their maximum values. It can also be defined as when the lagging edge of the input pulse and the lagging edge of the output pulse have both decayed to 50% of their maximum value. The propagation delay may impose serious limitations on the speed of the operation of the gate.

Figure 35.5 shows the result of a **PSpice** analysis of the propagation delay between the leading edge of a **TTL** input pulse and the leading edge of an output pulse of the 7804 gate.

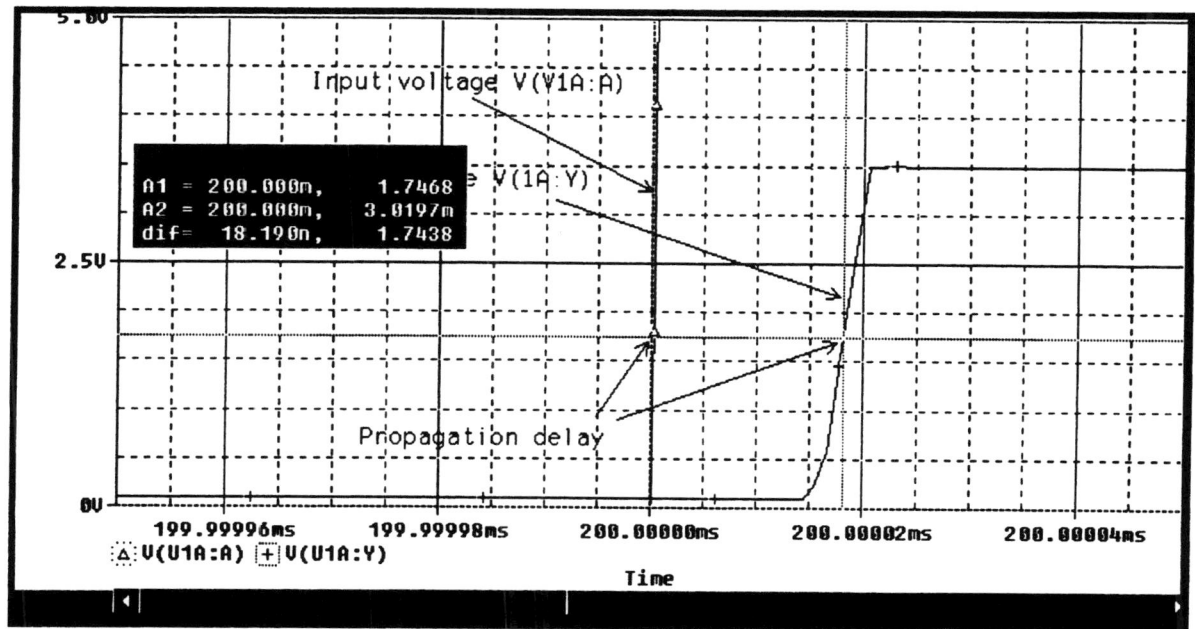

Figure 35-5

At 1.7468 volts, which is approximately 50% of the maximum output voltage, the propagation delay is measured at 18.19 nanoseconds.

    a. Determine the propagation delay at the lagging edge of a **TTL** pulse and the lagging edge of an output voltage pulse of the gate using the **PSpice** program.

    b. Determine the propagation delays both at the leading and lagging edges of these two voltages using experimental procedures. You may need to expand the time axis of the scope by using its 10X multiplier.

    c. Compare the results of the **PSpice** simulation with those obtained from the laboratory data for the leading edge deviation using the preceding percent deviation formula. The propagation delay measured by **PSpice** is the standard for comparison.

% deviation = _____

Part 5. NOT-AND Logic

Computer Simulation:

The circuit shown in Fig. 35.6 consists of an **AND** gate together with an **INVERTER** gate labeled "7404." We shall demonstrate both through a computer simulation and experimental data that the output of the **INVERTER** changes the logical and physical **ON** of the **AND** gate into a logical and physical **OFF**, and the logical and physical **OFF** of the **AND** gate into a logical and physical **ON**. The **INVERTER** gate is identified by the bubble at its output terminal.

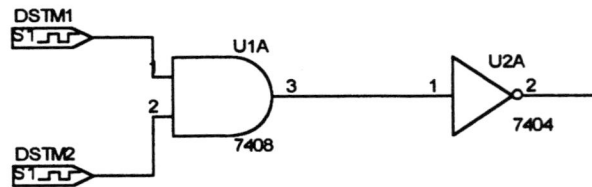

Figure 35-6

We shall keep the logic states at the input terminals of this system the same as we did for the **AND** gate in Part 1. Their **PROBE** traces are shown in Fig. 35.7. The objective is to obtain the logical relationship between the input terminals of the **AND** gate and the output terminal of the **INVERTER** gate. Construct Table 35.2. Note that the **INVERTER** input is also the output of the **AND** gate.

TABLE 35.2

Input1 (7408)	Input2 (7408)	Input1 (7404)	Output (7404)
1	1	1	
0	1	0	
1	0	0	
0	0	0	

Run a 500-millisecond **PSpice** simulation to obtain the logic states at the output terminal of the 7404 **INVERTER** gate. Its **PROBE** plot is labeled U2A:Y.

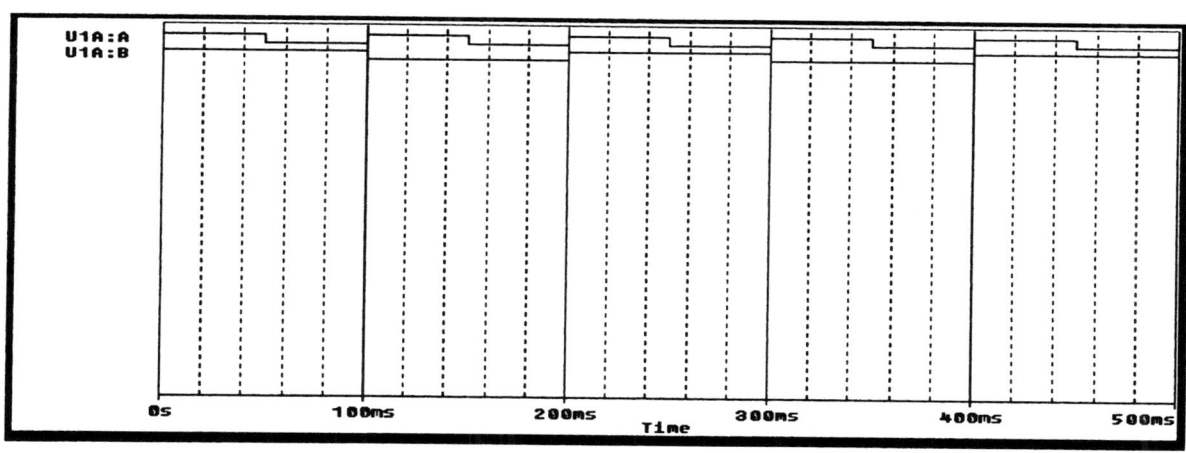

Figure 35-7

a. Obtain the logic states for the output terminal of the 7404 gate in Table 35.2.

b. What logical relationship exists between the input terminals of the **AND** gate and the output terminal of the **INVERTER** gate?

c. What is the most common logic state of the output terminal of the 7404 gate?

d. What are the logic states of the gates at 25 ms, 125 ms, and 375 ms?

Hint: Move a **PROBE** cursor to the indicated times. The logic state of each gate appears on the left margin of the **PROBE** plot.

Experimental Determination of Logic States:

Proceed as follows:

1. Plug the 7408 and 7404 gates into the protoboard of the Cadet Trainer. Wire the gates as shown.
2. Wire pin 14 to +5 volts and pin 7 to ground (zero volts).
3. Connect the **TTL** clock of the trainer to Input terminal 1 of the 7408 gate. This provides a train of pulses. Their fundamental frequency can be set.
4. Set the fundamental frequency of the **TTL** clock to 10 Hz.
5. Connect Input terminal 2 of the 7408 gate to the **HI/LO** logic switch of the trainer.
6. Connect the output terminal of the 7404 gate to a logic monitor or the logic probe. It can also be wired to an **LED**, if available, of the logic trainer. This gives a direct visual indication of the **ON** or **OFF** state of the **AND** gate.
7. Connect channel 1 of the scope to monitor the **TTL** clock.
8. Connect channel 2 of the scope to monitor at the output terminal of the 7404 gate.
9. Turn on the power to the trainer, and obtain the data.

From the experimental data, answer the following questions:

a. Are the experimentally determined logic states of the gates the same as those determined by the simulation?

b. What is the logical relationship between the output terminal of the 7408 gate and the input terminal of the 7404 gate?

c. What is the logical relationship between the output terminal of the 7408 gate and the output terminal of the 7404 gate?

d. Is the same result of the previous question obtained with the simulation data?

Part 6. 7400 NAND Gate

Computer Simulation:

The 7400 **NAND** gate shown in Fig. 35.8 performs the same logic function as the combination of the 7408 and the 7404 gates. Verify this using a **PSpice** simulation and an experimental determination of its logic states.

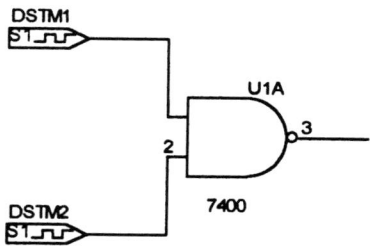

Figure 35-8

Run a 500-millisecond **PSpice** simulation to obtain the logic states of the input and output terminals of the 7400 gate. The **PROBE** plot of the output terminal is labeled U1A:Y.

a. From this simulation, complete the data entry into Table 35.3.

TABLE 35.3

Input terminal 1	Input terminal 2	Output terminal 3
1	1	
0	1	
1	0	
0	0	

b. Sketch the logic states of the output terminal on the **PROBE** plot shown in Fig. 35.9.

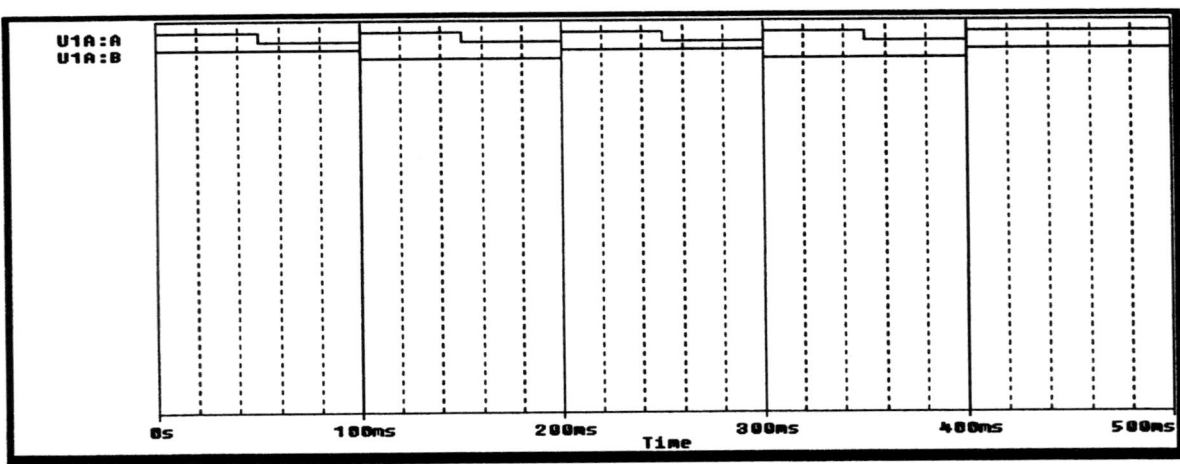

Figure 35-9

Experimental Determination of Logic States:

a. Replace the components of Fig. 35.6 on the protoboard with the 7400 gate of Fig. 35.8. Proceed with the experiment as in Part 5. After the completion of the experiment, enter the logic states of the output terminal in Table 35.4.

TABLE 35.4

Input terminal 1	Input terminal 2	Output terminal 3
1	1	
0	1	
1	0	
0	0	

b. Compare the logic states of Output terminal 3 in Table 35.4 with those of Output terminal 3 in Table 35.3. Are they the same?

Name _____
Date _____
Instructor _____

EXPERIMENT 36

Analysis of OR, NOR, and XOR Logic Gates

OBJECTIVES

1. To analyze logical primitives using the **PSpice** program.
2. To construct circuits in the laboratory using these logical primitives.
3. To compare the results obtained in the laboratory with those obtained from simulation.
4. To make any needed corrections in the circuits.

EQUIPMENT REQUIRED

Instruments

Computer able to run **OrCad PSpice**, version 9.1 or later
CADET logic trainer (or equivalent)
Tektronic 2236 oscilloscope (or equivalent)
10:1 scope probe
DMM
7402, 7408, 7432, 7486 gates

EQUIPMENT ISSUED

Item	Laboratory serial no.
Logic trainer	
Oscilloscope	
Scope probe	

RÉSUMÉ OF THEORY

We shall next investigate logic operations of the **OR, NOR,** and **XOR** gates. Also, we shall combine the previously studied **AND** gate with the **OR** gate and obtain the overall logic of that combination.

PROCEDURE

For this experiment, the same procedures used in Experiment 35 are followed. They are repeated here for ease of reference. The student is to design and simulate all logical circuits using the **PSpice** program before he/she performs the actual experiment, which conforms to engineering practice. The result of the simulation can then be discussed with the laboratory instructor, preferably in the week prior to a particular experiment.

After completion of the simulation, the design is taken into the laboratory. After the actual circuit is constructed and measurements are taken, the student should continuously compare the results obtained from the simulation with those obtained in the laboratory. In this way, any discrepancy between the results can be cleared up.

Part 1. OR Gate: Computer Simulation

Figure 36.1 shows a two-input **OR** gate. The **OR** gate is located in the **eval.slb** library of the **PSpice** program. Its shape tells its logical function. The circuit consists of the 7432 logic gate. Digital stimulus generator **DSTM1** is a digital clock connected to Input terminal 1 of the gate. It generates a stream of pulses analogous to the **TTL** pulse train. Its pulse parameters are specified as shown. Double-click on each of them to set them. In the **Display Property** box, type in the desired values. **DSTM2** is connected to Input terminal 2 of the gate. The logic states of the **DSTM2** are set by the **OrCad Capture-[Property Editor]**. Both stimulus generators are located in the **Source** library of the **PSpice** program. Terminal 3 of the 7432 gate is the output terminal.

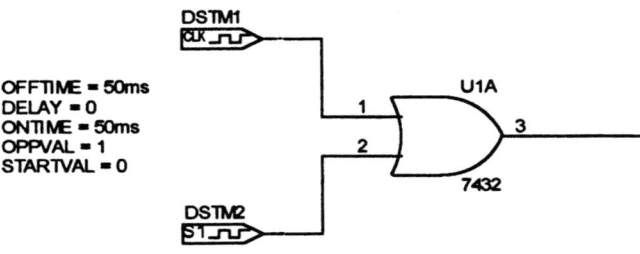

Figure 36-1

Our objective is to define the logic function of the **OR** gate.

As in the previous experiment, each of the stimulus generators can assume two states. Logically, they are known as 1 or **On**, and 0 or **Off**. Thus, there are four logic states at the input terminals. The **OR** gate responds to each of those. The conditions specified at the input terminals are logic states, not electrical quantities such as voltage. To keep track of the possible logic states, construct Table 36.1.

TABLE 36.1

Input terminal 1	Input terminal 2	Output terminal 3
1	1	
0	1	
1	0	
0	0	

In this table, all possible combinations of logic input states of the input terminals are present.

Next, run a 500-millisecond **PSpice** simulation to obtain the logic states of Output terminal 3.

Set the logic states at the input terminals as shown on the **PROBE** plot (Fig. 36.2). The trace of Input terminal 1 is shown as **U1A:A**. Its frequency is set at 10 Hz. The trace of Input terminal 2 is shown as **U1A:B**. Its frequency is set at 5 Hz. These frequencies are set relatively low to allow the experimenter to observe the logic states if an **LED** is used during the experiment.

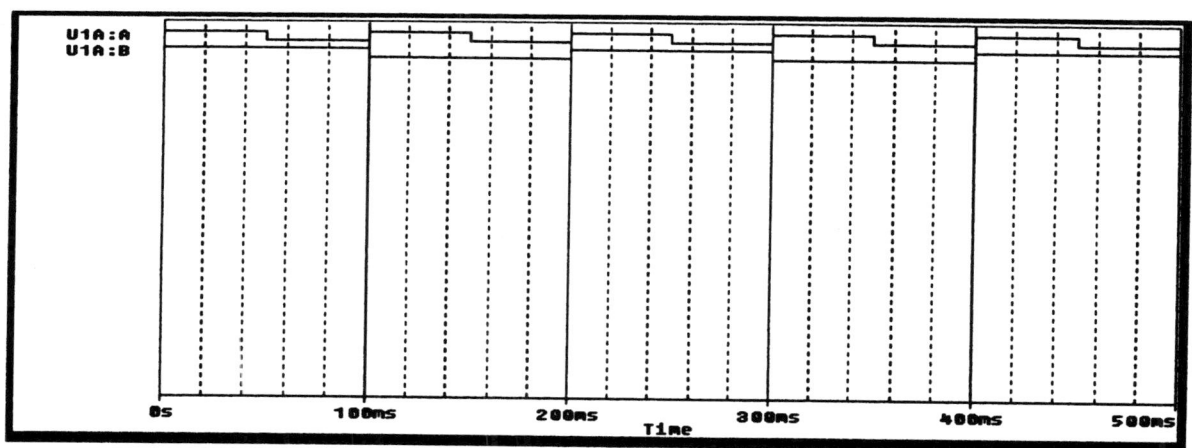

Figure 36-2

The **PROBE** plot shows that the logic states of the input terminals are identical to those listed in Table 36.1. After the logic states of the input terminals are set, we perform the simulation of Figure 36.1. In addition to the two traces in the **PROBE** plot, we now obtain the logic state of Output terminal 3. Its trace is plotted as **U1A:Y**. The letters **A** and **B** are always used to define input terminals. The letter **Y** is used to define the output terminal.

From the simulation data obtained, answer the following:

    a. Complete the logic states for Output terminal 3 in Table 36.1, and sketch their trace in the preceding **PROBE** plot.

    b. What logical relationship exists between the input terminals and the output terminal?

    c. What is the most common logic state of Output terminal 3? Compare this to the most common logic state of the output terminal of the 7408 gate.

    d. What are the logical states of the 7432 **OR** gate at 25 ms, 125 ms, and 375 ms? *Hint:* Move a **PROBE** cursor to the indicated times. The logic state of each terminal will appear on the left margin of the **PROBE** plot.

Part 2. OR Gate: Experimental Determination of Logic States

This determination is performed using the Cadet Trainer of E & L Instruments (or its equivalent) and the Tektronic 2236 oscilloscope.

Proceed as follows:

1. Plug the 7432 gate into the protoboard of the Cadet Trainer.
2. Wire pin 14 to +5 volts and pin 7 to ground (zero volts).
3. Connect the **TTL** clock of the trainer to Input terminal 1. This provides a train of pulses. Their fundamental frequency can be set.
4. Set the fundamental frequency of the **TTL** clock to 10 Hz.
5. Connect Input terminal 2 of the gate to the **HI/LO** logic switch of the trainer.
6. Connect Output terminal 3 of the gate to a logic monitor or the logic probe. It can also be wired to an LED of the logic trainer. This gives a direct visual indication of the **ON** or **OFF** state of the **OR** gate.
7. Connect channel 1 of the scope to monitor the **TTL** clock.
8. Connect channel 2 of the scope to monitor Output terminal 3.
9. Turn on the power to the trainer.
10. Toggle the **HI/LO** logic switch to obtain the logic state of the gate.

From the experimental data obtained, answer the following questions:

a. How does the period of the pulse of the **TTL** pulse train compare to the period of the **PROBE** trace of **U1A:A**?

b. What is the frequency of the trace and that of the **TTL** pulses?

c. Are the logic states of the 7432 the same in both the computer simulation and in the experiment as measured at 25 ms, 125 ms, and 375 ms?

d. What was the voltage amplitude of the **TTL** pulses and that of Output terminal 3?

e. If there was a difference in their respective voltages, how can this be explained?

Part 3. Logic States versus Voltage Levels

The **PSpice Probe** plot provided traces of the logic states of the three terminals of the 7432 gate. The experimental data provided two distinct voltage levels at these terminals corresponding to the logic states. The high voltage output of the 7432 gate corresponded to a logical 1, or **ON**, on the **PROBE** plot. The low voltage output of that gate corresponded to a logical 0, or **OFF**, on the **PROBE** plot. We can simulate the experimental voltage levels using the **PSpice** program.

First, modify the logic circuit of Fig. 36.1. It is shown as Fig. 36.3.

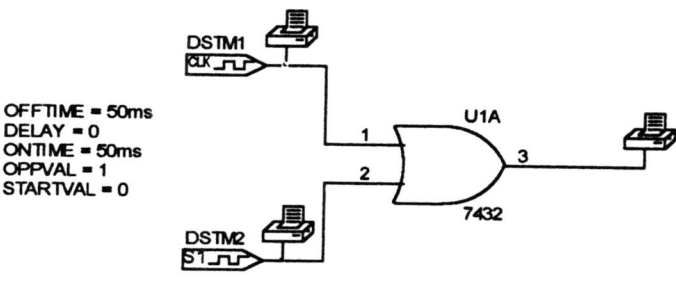

Figure 36-3

To each terminal, a **VPrint1** device is attached. It can be found in the **Special** library of the **PSpice** program. This enables you to print the voltages of the 7432 **OR** gate. Run a simulation.

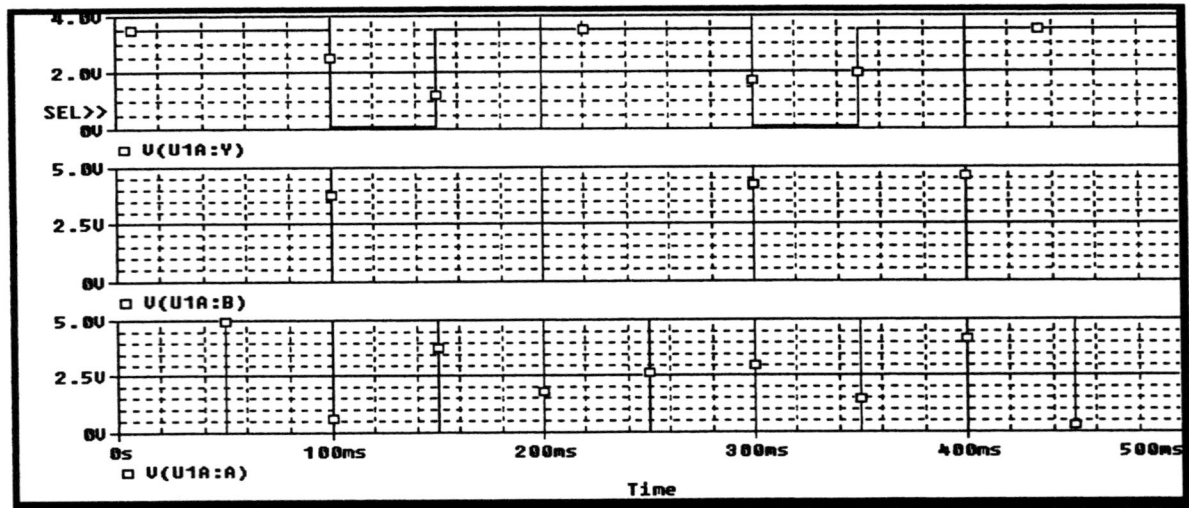

Figure 36-4

The results are shown on the **PROBE** plot in Fig. 36.4. It shows that the two input voltages are 5 volts at maximum. The output voltage has a maximum value of 3.5 volts. The traces now have the letter **V** (for voltage) attached to them as they did for the 7408 gate. The terminal designations **A**, **B**, and **Y** are also retained as for the 7408 gate.

    a. Calculate the percent deviation between the experimentally obtained output voltage VY and the simulated output voltage shown on the **PROBE** plot as trace V(V1A:Y). Use the following formula:

$$\% \text{ deviation} = \frac{V(V1A{:}Y) - VY}{V(V1A{:}Y)} * 100\%$$

% deviation = _____

In this calculation, V(V1A:Y) is the standard of comparison.

    b. If a 10% deviation between the simulated voltage V(V1A:Y) and the experimental voltage VY is acceptable, does the calculated percent deviation fall within the permissible range?

    c. How does this percent deviation compare to that of the 7408 gate?

Part 4. Combining AND with OR Logic

Computer Simulation:

The circuit shown in Figure 36.5 consists of two **AND** gates, **U1A** and **U2A**, connected to an **OR** gate, **U3A**.

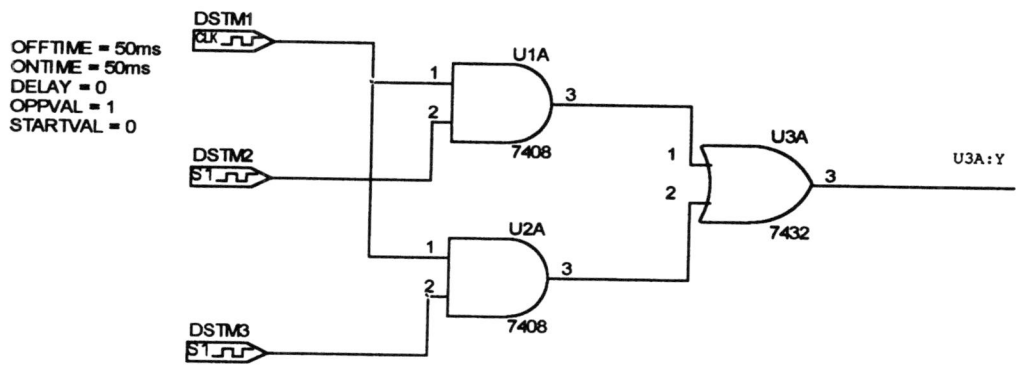

Figure 36-5

Input terminals 1 of the **U1A AND** and **U2A AND** gates are connected to a digital clock. Its parameters are set as shown. Its trace, **DSTM1**, is shown on the **PROBE** plot in Fig. 36.6. Input terminals 2 of the two **AND** gates are connected to **DSTM2** and **DSTM3**, respectively. Their traces are shown as **TM2:pin1** and **TM3:pin1**, respectively, on the **PROBE** plot. The logic state of **DSTM3** is a logical **HIGH** for the duration of the simulation as shown on its trace. The output terminals of the two **AND** gates are connected to the two input terminals of the **OR** gate. Terminal 3 of the 7432 gate is the output terminal of the **OR** gate.

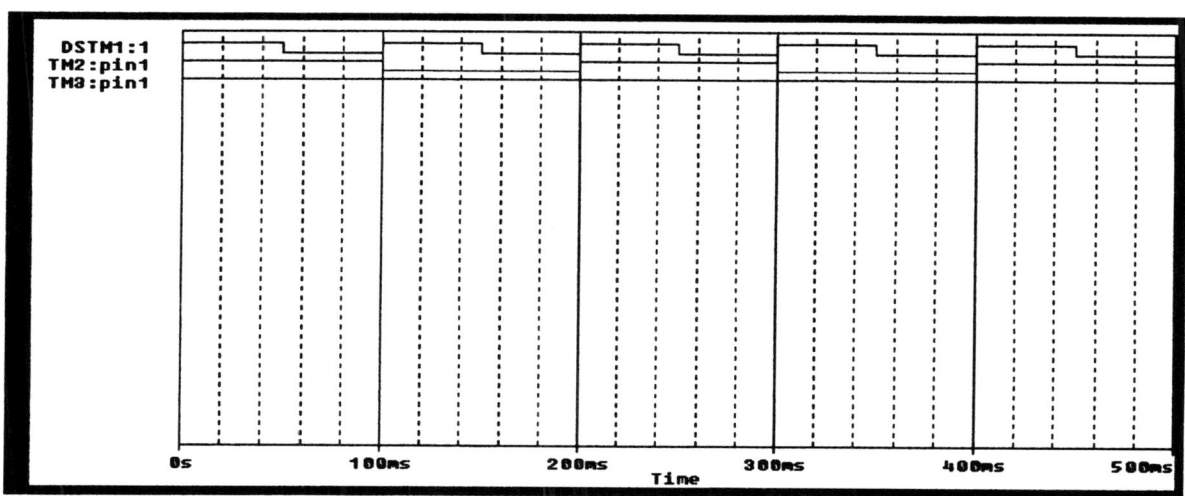

Figure 36-6

Run a 500-millisecond **PSpice** simulation to obtain the requested data in Table 36.2.

a. At the completion of the simulation, obtain a **PROBE** plot, which contains all nine traces of the three logic gates.

b. Using this information, complete Table 36.2.

TABLE 36.2

U1A:A	U1A:B	U1A:Y	U2A:A	U2A:B	U2A:Y	U3A:A	U3A:B	U3A:Y
1	1		1	1				
0	1		0	1				
1	0		1	1				
0	0		0	1				

c. What are the logic states of the **U3A:Y** terminal at 25 ms, 125 ms, and 375 ms?

d. During the 500-ms duration of the simulation, is the **U3A:Y** gate mostly in the **ON** state or the **OFF** state?

Experimental Determination of Logic States

Proceed as follows:

1. Plug the 7408 and 7432 gates into the protoboard of the Cadet Trainer as shown in Fig. 36.5. Wire the gates as shown.
2. Wire pin 14 to +5 volts and pin 7 to ground (zero volts).
3. Connect the **TTL** clock of the trainer to Input terminals 1 of the two 7408 **AND** gates. This provides a train of pulses. Their fundamental frequency can be set.
4. Set the fundamental frequency of the **TTL** clock to 10 Hz.
5. Connect Input terminal 2 of the **U1A AND** gate to the **HI/LO** logic switch of the trainer.
6. Connect Input terminal 2 of the **U2A AND** gate to a constant 5-volt supply of the trainer.
7. Connect the output terminal of the 7432 **OR** gate to a logic monitor or the logic probe. It can also be wired to an **LED** of the logic trainer. This gives a direct visual indication of the **ON** or **OFF** state of the **OR** gate.
8. Connect channel 1 of the scope to monitor the **TTL** clock.
9. Connect channel 2 of the scope to monitor at the output terminal of the 7432 gate.

10. Turn on the power to the trainer.
11. Toggle the **HI/LO** logic switch to obtain the logic state of Output terminal 3 of the 7432 **OR** gate.

From the experimental data, answer the following questions:

a. Are the experimentally determined logic states of the gates the same as those determined by the simulation?

b. What is the logical relationship between the output terminal of the 7432 **OR** gate and the **TTL** clock?

c. What is the logical relationship between the output terminals of the **AND** gates and the output terminal of the **OR** gate?

Part 5. NOR and XOR Logic Combined

Computer Simulation:

This section introduces the **NOR** and **XOR** gates. **NOR** is a contraction of **NOT-OR**, and **XOR** is a contraction of **EXCLUSIVE-OR**. Figure 36.7 contains both types of gates that perform these logic functions. The **NOR** gate **U1A** is identified by a bubble at its output terminal and the **XOR** gate **U2A** by the added line across its input terminals. The clock parameters of the **DSTM1** are set as shown. The **DSTM2** frequency is set at 5 Hz. Their traces—the result of a 500-millisecond simulation—are shown in Fig. 36.8.

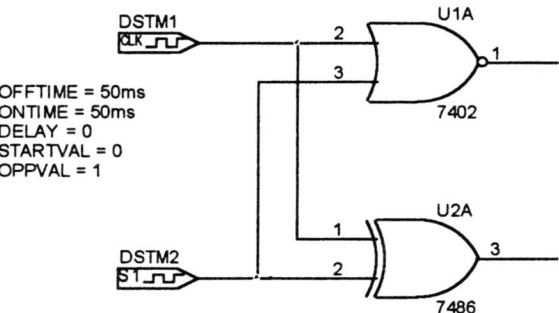

Figure 36-7

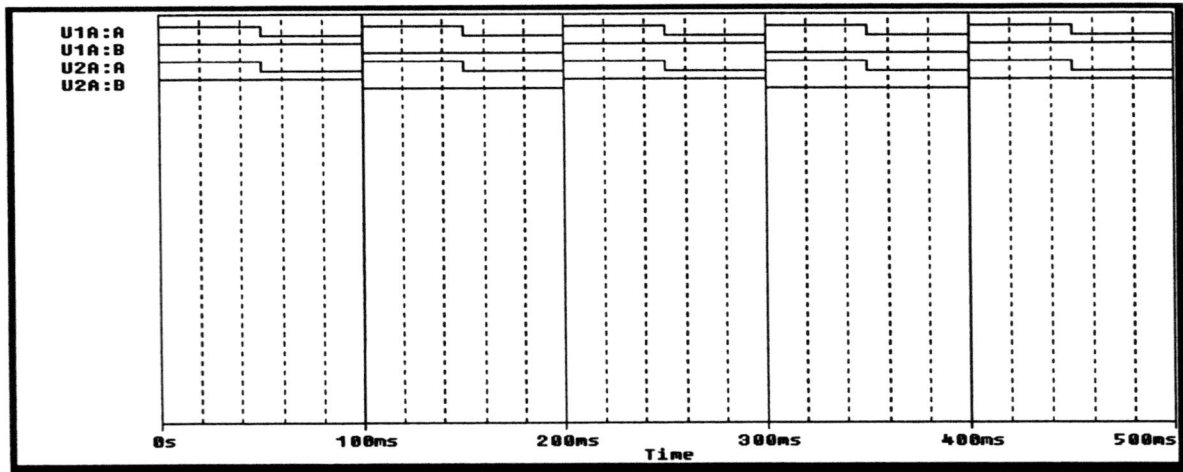

Figure 36-8

The inputs to terminals **U1A:A** and **U2A:A** are identical because both terminals are connected to the clock. The inputs to terminals **U1A:B** and **U2A:B** are identical because both terminals are connected to **DSTM2**.

  a. Run a 500-millisecond simulation, and obtain the **PROBE** traces for the output terminal of the **NOR** and **XOR** gates.

  b. Complete all data entries in Table 36.3.

TABLE 36.3

U1A:A	U1A:B	U1A:Y	U2A:A	U2A:B	U2A:Y
1	1		1	1	
0	1		0	1	
1	0		1	0	
0	0		0	0	

  c. From the data in Table 36.3, what is the logic of the **NOR** gate?

  d. How does this compare to the logic of the **OR** gate?

  e. From the data in Table 36.3, what is the logic of the **XOR** gate?

f. How does this compare to the logic of the **OR** gate?

Experimental Determination of Logic States:

Proceed as follows:

1. Plug the 7402 and 7486 gates into the protoboard of the Cadet Trainer. Wire the gates as shown in Fig. 36.7.
2. Wire pin 14 to +5 volts and pin 7 to ground (zero volts).
3. Connect the **TTL** clock of the trainer to Input terminal 2 of the 7402 **OR** gate and Input terminal 1 of the 7486 **XOR** gate.
4. Set the fundamental frequency of the **TTL** clock to 10 Hz.
5. Connect Input terminal 3 of the 7402 **OR** gate and Input terminal 2 of the **XOR** gate to the **HI/LO** logic switch of the trainer.
6. Connect the output terminal of the 7402 **NOR** gate to a logic monitor or the logic probe. It can also be wired to an **LED** of the logic trainer. This gives a direct visual indication of the **ON** or **OFF** state of the **NOR** gate.
7. Repeat step 6 for the output terminal of the 7486 **XOR** gate. This gives a direct visual indication of the **ON** or **OFF** state of the **XOR** gate.
8. Connect channel 1 of the scope to monitor the **TTL** clock.
9. Connect channel 2 of the scope alternately to the output terminals of the two gates.
10. Turn on the power to the trainer.
11. Toggle the **HI/LO** logic switch to obtain the logic states of the output terminals of the two gates.

From the experimental data, answer the following questions:

a. Are the experimentally determined logic states of the gates the same as those determined by the simulation? Compare your experimental data with that contained in Table 36.3.

b. What is the logical relationship between the output terminal of the 7402 **OR** gate and the **TTL** clock?

c. What is the logical relationship between the output terminal of the 7402 **OR** gate and its input terminals? Compare your answer with the data in Table 36.3.

d. What is the logical relationship between the output terminal of the 7486 **XOR** gate and the **TTL** clock?

e. What is the logical relationship between the output terminal of the 7486 **XOR** gate and its input terminals? Compare your answer with the data in Table 36.3.

f. How does the logic of the **OR** gate differ from that of the **XOR** gate?

g. Define the term **EXCLUSIVE-OR**.

Name _____
Date _____
Instructor _____

Analysis of Integrated Circuits

EXPERIMENT **37**

OBJECTIVES

1. To analyze various integrated circuits, or chips, using the **PSpice** program.
2. To construct circuits in the laboratory using chips.
3. To compare the results obtained in the laboratory with those obtained from simulation.
4. To make any needed corrections in the circuits.

EQUIPMENT REQUIRED

Instruments

Computer able to run **OrCad PSpice**, version 9.1 or later
Cadet logic trainer (or equivalent)
Tektronic 2236 oscilloscope (or equivalent)
10:1 scope probe
DMM
7493A, 7474, 74107 chips

EQUIPMENT ISSUED

Item	Laboratory serial no.
Logic trainer	
Oscilloscope	
Scope probe	

461

RÉSUMÉ OF THEORY

In this experiment, we extend our investigation into systems containing several of the logical primitives investigated in the previous two experiments. The components of these systems are combined electrically and mounted together in circuits known as *integrated circuits* or *chips*. These integrated circuits perform an overall function to which each of their components contributes. Basically, these functions fall into two broad categories: decision-making logic circuits and memory circuits. The former are known as *gates* and the latter as flip-flop latches. In our analysis, we take a macroscopic approach; that is, we concentrate on their terminal behavior without analyzing the behavior of individual components.

Today, there is a bewildering number of integrated circuits in existence; new ones continue to be designed to serve new functions. This laboratory is an introduction to these devices.

PROCEDURE

Part 1. Positive Edge Triggered D Flip-Flop

Computer Simulation:

Figure 37.1 depicts an integrated circuit, known as the *7474 positive edge triggered D flip-flop*. The term "positive edge triggered" means that if there is a logical **HIGH** on the **D** input terminal and if the clock input **DSTM1** connected to the **CLK** input terminal goes positive, the flip-flop will **SET**. It will be at a logical **ON**. Otherwise, the flip-flop is **RESET** and is at a logical **OFF**.

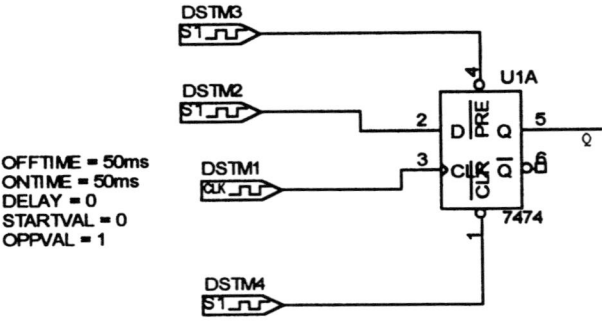

Figure 37-1

The 7474 flip-flop has four input terminals. For positive edge triggering, the **PRE** and **CLR** input terminals must be at a logical **ON**. Thus, **DSTM3** and **DSTM4** are held at a logical **ON** throughout the experiment. **DSTM1** is the clock input connected to the **CLK** input terminal. Its parameters are set as shown. **DSTM2** is connected to the **D** input terminal. It is set to a frequency of 6 Hz. Their traces are shown on the **PROBE** plot in Fig. 37.2.

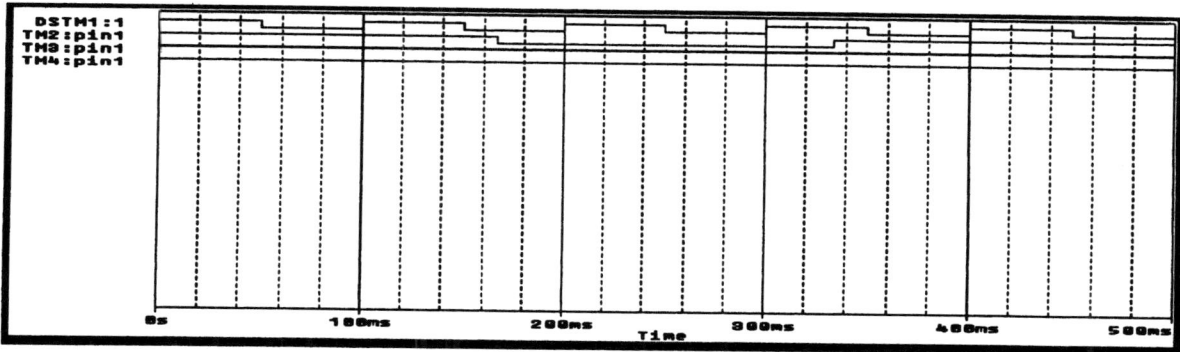

Figure 37-2

Initial State of the Flip-Flop

At the beginning of the simulation, three possible logic states exist for the flip-flop. Depending on its prior history, the flip-flop can be either in the **RESET** or **OFF** state, the **SET** or **ON** state, or an indeterminate state, symbolized as **X** by the **PSpice** program. The desired initial state of the flip-flop is set as follows:

1. Click on the **PSpice** tab. The New Simulation box opens.
2. Enter a number into the Name box. If this is the first simulation of a particular circuit, use the number 1.
3. Click on Create. The Simulation settings box opens.
4. Click on the Option tab.
5. In the Category box, click on Gate-level Simulation.
6. Click in Initialize all flip-flops to.
7. Click on 1 to select it. This sets the initial state of the flip-flop to **ON**.

In this example, set the initial state of the 7474 flip-flop to **SET** or **ON**.

Run a 500-millisecond **PSpice** simulation to obtain the logic states of the output terminal **Q** of the flip-flop.

From the simulation data obtained, answer the following:

  a. Is the initial state of the flip-flop in the **SET** or **ON** condition?

  b. At what time does the initial state of the flip-flop change? Why?

  c. What is the condition of the **D** input when the **Q** output is low?

d. At what time does the flip-flop return to its **SET** condition? Why?

e. Write a general statement that relates the logic states of output of the flip-flop to the logic states of its input terminal.

f. Relate your answer in the previous statement to the first paragraph of Part 1.

g. In your own words, from the simulation data obtained, define a positive edge triggered D flip-flop.

h. Modify the flip-flop in Fig. 37.1 by adding **VPLOT1** devices to each of its inputs and the output terminal. Make a note of the voltage levels obtained, which will be used as a reference for the experimental data.

i. How does the flip-flop function as a memory device?

Experimental Determination of Logic States:

Proceed as follows:

1. Plug the 7474 flip-flop into the protoboard of the Cadet Trainer.
2. Wire pin 14 to +5 volts and pin 7 to ground (zero volts).
3. Connect the **TTL** clock of the trainer to Input terminal 3, the **CLK** terminal.
4. Set the fundamental frequency of the **TTL** clock to 10 Hz.
5. Connect Input terminals **PRE** and **CLR** to a 5-volt supply to keep both of them at a logical **HIGH**.
6. Connect Input terminal 2, the **D** terminal of the flip-flop, to the **HI/LO** logic switch of the trainer.
7. Connect channel 1 of the scope to monitor the **TTL** clock.
8. Connect channel 2 of the scope to monitor Output terminal 5, labeled **Q**.
9. Connect Output terminal 5 to a logic monitor or a logic probe. It can also be wired to an **LED** of the logic trainer.
10. Turn on the power to the trainer.
11. Toggle the **HI/LO** logic switch, and obtain the logic state of the flip-flop.

From the experimental data obtained, answer the following questions:

a. Are the **PRE** and **CLR** terminals held **HIGH** at 5 volts during the experiment?

b. What is the voltage of the **TTL** pulse and the input pulse at the **D** terminal?

c. What is the output voltage at the **Q** terminal?

d. Explain the difference in amplitude between the input voltages and the output voltage.

e. Compare the logical transition states of the output voltage **Q** from the experimental data with those obtained from the simulation. Do they occur at the same time?

Part 2. Frequency Division

Computer Simulation:

The circuit shown in Fig. 37.3 contains two negatively edge triggered flip-flops. Each of its output terminals **Q** will have a frequency one-half that of its **CLK** input. Thus, we can expect that the frequency of the terminal **U2A:Q** will be one-fourth that of the input terminal **U1A:CLK**.

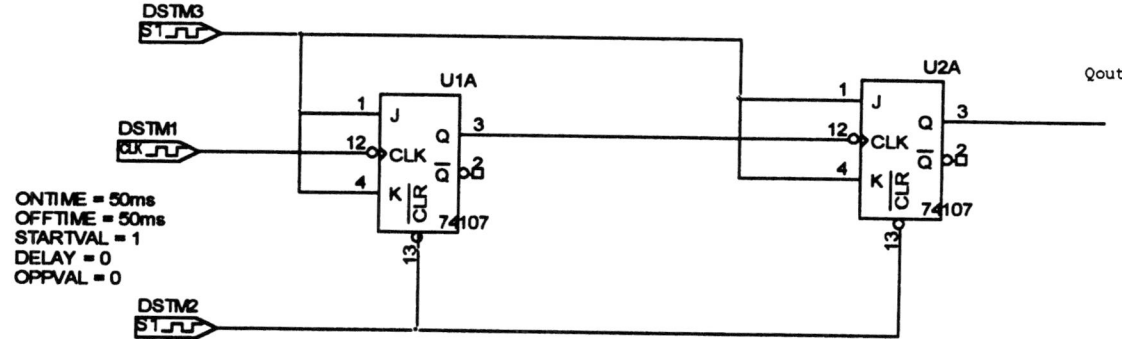

Figure 37-3

The clock input, **DSTM1**, is set with the parameters as shown. **DSTM2** and **DSTM3** are both held at a logical **HIGH**. Set the initial condition of the flip-flops to 0, or **RESET**, by the method previously stated. The stimulus conditions are shown on the **PROBE** plot in Fig. 37.4.

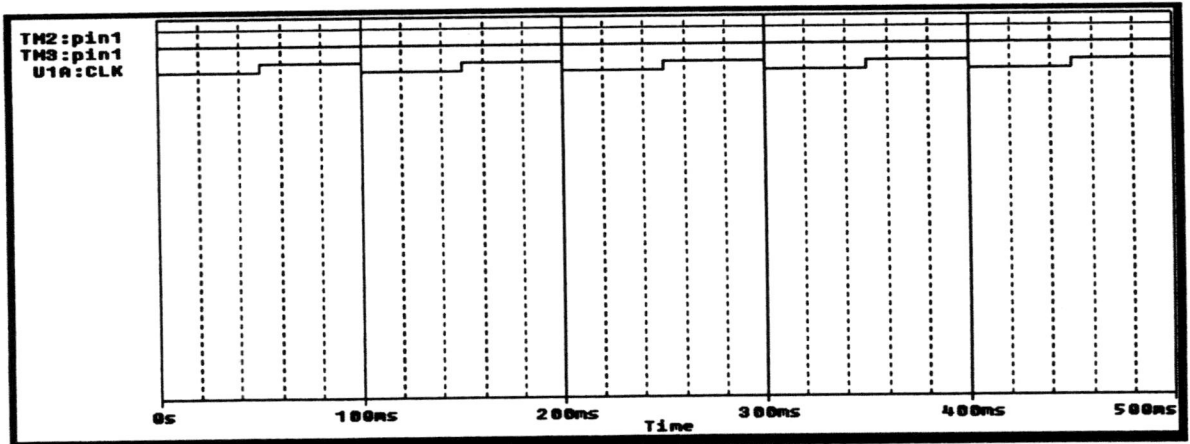

Figure 37-4

Run a 500-millisecond **PSpice** simulation to obtain the logic state of the output terminals **U1A:Q** and **U2A:Q** of the two flip-flops.

From the simulation data obtained, answer the following:

    a. What is the frequency of the output pulse at terminal **U1A:Q**?

    b. How does its frequency compare to that of the input pulse **U1A:CLK**?

    c. What is the frequency of the output pulse at terminal **U2A:Q**?

    d. How does its frequency compare to that of the input pulse **U2A:CLK**?

    e. Compare the overall frequency reduction of the output pulse **U2A:Q** to the input pulse **U1A:CLK**.

f. What frequency reduction was achieved by each of the 74107 flip-flops?

g. How many 74107 flip-flops would be required to achieve a frequency reduction of ten between the output terminal **Q** of the last flip-flop and the terminal **U1A:CLK**?

Experimental Determination of Logic States:

Proceed as follows:

1. Plug the two 74107 flip-flops into the protoboard of the Cadet Trainer and wire them as shown.
2. Wire pin 14 to +5 volts and pin 7 to ground (zero volts).
3. Connect the **TTL** clock of the trainer to Input terminal **U1A:CLK**.
4. Set the fundamental frequency of the **TTL** clock to 10 Hz.
5. Connect Input terminals **U1A:J** and **U2A:J** together to a 5-volt supply to keep both of them at a logical **HIGH**.
6. Connect Input terminals **U1A:CLR** and **U2A:CLR** together to a 5-volt supply to keep them both at a logical **HIGH**.
7. Connect channel 1 of the scope to monitor the **TTL** clock.
8. Connect channel 2 of the scope to monitor Output terminal **U2A:Q**.
9. Also, connect Output terminal **U2A:Q** to a logic monitor or a logic probe. It can also be wired to an **LED** of the logic trainer.
10. Turn on the power to the trainer, and obtain the data.

From the experimental data obtained, answer the following questions:

a. Verify that the **J** and **CLR** terminals of the two flip-flops are held high at 5 volts during the experiment.

b. What are the voltage levels at the **U1A:CLK**, **U2A:CLK**, and **U2A:Q** terminals? If they are different, explain the reason for that difference.

c. Compare the periods of the **U1A:CLK**, **U2A:CLK**, and **U2A:Q** pulses from your oscilloscope data.

d. From that data, determine the frequency of these pulses.

e. How do they compare to the computer simulation data obtained?

Part 3. Asynchronous Counter: 7493A Integrated Circuit

Computer Simulation:

The 7493A chip consists of four **JK** flip-flops and an **AND** gate. It is defined as an asynchronous counter because not all pulse transitions within the counter occur in synchronism with the clock pulse. This is because there is an inherent propagation delay as a pulse traverses a flip-flop. The circuit shown in Fig. 37.5 is wired as a **MOD-10** counter.

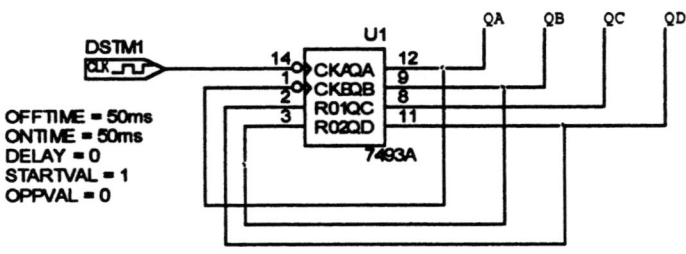

Figure 37-5

The parameters of the clock pulse **DSTM1** are set as shown. The initial state of the 7403 device was set to zero volts.

a. Perform a two-second **PSpice** analysis.

b. Obtain a **PROBE** plot containing the traces of **QA, QB, QC, QD**, and the clock pulse of **DSTM1**.

c. Place the cursor at 20 milliseconds to confirm the initial state of the device.

d. Place the cursor at 175 milliseconds. Compare the number of clock pulses to the left of the cursor with the binary number indicated by the logic state of the **Q** terminals. Are they the same?

e. Place the cursor at 375 milliseconds. Compare the number of clock pulses to the left of the cursor with the binary number indicated by the logic state of the **Q** terminals. Are they the same?

f. Place the cursor at 575 milliseconds. Compare the number of clock pulses to the left of the cursor with the binary number indicated by the logic state of the **Q** terminals. Are they the same?

g. Place the cursor at 1.075 seconds. Compare the number of clock pulses to the left of the cursor with the binary number indicated by the logic state of the **Q** terminals. Are they the same?

h. Are the initial states of all terminals the same as at the beginning of the counting process when the cursor is at 1.075 seconds?

i. From your finding in the previous question, define a **MOD-10** counter.

j. Which of the **Q** terminals represents the most significant digit? Which represents the least significant one?

k. Measure the propagation delay between the clock pulse of **DSTM1** and the pulse at the output terminal **QA**. *Hint:* change the time axis of the **PROBE** plot to a suitable scale.

Experimental Determination of Logic States:

Proceed as follows:

1. Plug the two 7493A counters into the protoboard of the Cadet Trainer. Wire it as shown.
2. Wire pin 14 to +5 volts and pin 7 to ground (zero volts).
3. Connect the **TTL** clock of the trainer to Input terminal **U1:CKA**.
4. Set the fundamental frequency of the **TTL** clock to 10 Hz.
5. Connect terminal **CKB** to the **QA** output terminal.
6. Connect terminal **RO1** to the **QD** output terminal.
7. Connect terminal **RO2** to the **QB** output terminal.
8. The **QC** output terminal is at pin 8.
9. Connect channel 1 of the scope to monitor the **TTL** clock. Set the time axis of the scope to obtain at least ten cycles of the TTL pulse.

10. Connect channel 2 of the scope to monitor Output terminals **QA**, **QB**, **QC**, and **QD**.
11. Connect the output terminals to a logic monitor or a logic probe. They can also be wired to an **LED** of the logic trainer.
12. Turn on the power to the trainer, and obtain the data.

From the experimental data obtained, answer the following questions:

   a. Compare the logic states of the output terminals after 1 pulse, 5 pulses, and 10 pulses of the **TTL** pulse.

   b. Compare this data with that obtained from the **PSpice** simulation.

   c. Measure the propagation delay between the **TTL** pulse and the trace of the **QA** terminal. Be sure to expand the time axis by the multiplier for the time axis.

   d. Does your experimental data of the measured propagation delay agree with the computer simulation data?